Imms' Outlines
of Entomology

IMMS' OUTLINES
OF ENTOMOLOGY

O. W. RICHARDS, F.R.S.

Emeritus Professor of Zoology and Applied Entomology,
Imperial College, University of London

AND

R. G. DAVIES

Reader in Entomology, Imperial College,
University of London

SIXTH EDITION

CHAPMAN AND HALL

LONDON and NEW YORK

First published 1942
by Methuen and Co Ltd
Second edition 1944
Third edition 1947
Fourth edition 1949
Fifth edition revised by
O. W. Richards & R. G. Davies 1959
Reprinted 1961
Reprinted with corrections 1967
Reprinted 1970
Reprinted 1973

Sixth edition published 1978
by Chapman and Hall Ltd
11 New Fetter Lane, London EC4P 4EE
Reprinted 1981

Published in the USA by
Chapman and Hall
in association with Methuen, Inc
733 Third Avenue, New York, NY 10017
© 1978 O. W. Richards & R. G. Davies

Printed in Great Britain by
Richard Clay (The Chaucer Press) Ltd
Bungay, Suffolk

ISBN 0 412 21660 4 (cased edition)
ISBN 0 412 21670 1 (paperback edition).

Contents

Preface to the Sixth Edition

In his preface to early editions of this book, the late Dr. A. D. Imms said that he intended it to be an elementary account of entomology as a branch of general biology. He had especially in mind the needs of university students of zoology and agriculture, as well as those intending later to specialize in entomology, and he suggested that the book might also interest teachers of advanced biology in schools.

These general aims and the balance between the different aspects of the subject have changed little in this and in our previous revision. We have, however, tried to bring the present edition up to date on the lines of our revised tenth edition of *Imms' General Textbook of Entomology*, published in 1977. The text has been entirely re-set and eleven illustrations have been replaced by new figures. The same orders of insects are recognized as in the last edition, but the sequence in which the Endopterygote groups appear has been changed to reflect more accurately their probable evolutionary relationships. Many small changes and some additions have been made in the physiological sections, the chapter on the origin and phylogeny of insects has been rewritten, and a new bibliography provides a selection of modern references for the intending specialist. It has been our object to make these alterations without materially increasing the length of the book or its level of difficulty.

We are greatly indebted to the authors on whose illustrations some of our diagrams are based; their names will be found in the legends to the figures concerned.

September, 1977   O.W.R.
                                                      R.G.D.

I

Introduction

Insects are segmented animals with a relatively tough integument and jointed limbs. They breathe by means of air-tubes or tracheae and the body is divided into head, thorax and abdomen. The head is the sensory and feeding centre, bearing the mouthparts and a single pair of antennae, perhaps homologous with the antennules of the Crustacea; compound eyes are usually present and often simple eyes or ocelli. The thorax is the locomotor centre, carrying three pairs of legs and usually two pairs of wings. The abdomen is the metabolic and reproductive centre; it contains the gonads and organs of digestion and excretion and usually bears special structures used in copulation and egg-laying. When it leaves the egg, the young insect differs more or less extensively from the adult form and its development therefore involves some degree of metamorphosis.

The number of known kinds, or species, of insects is difficult to estimate but certainly exceeds that of all other animals together. That approximately 800 000 different insects have been named and described is probably a conservative estimate. Usually several thousand new species are described in a single year, but notwithstanding this rate of discovery there is no doubt that the numbers yet to be brought to light exceed those of all the known kinds. The single order Coleoptera, or beetles, alone comprises over 330 000 named species. Even the one family Curculionidae, or weevils, includes more than 60 000 known species, while the Carabidae, or ground beetles, number about 25 000 kinds.

This remarkable capacity for differentiation shown by insects does not lend itself to exact analysis. It is, however, a matter of interest to consider some attributes which have most likely helped the members of this class to attain their dominant position in the animal kingdom.

(a) *Capacity for Flight.* The majority of insects are not wholly confined to the ground and vegetation but are also able to fly. The possession of wings provides unique means of dispersal, of discovering their mates, of seeking food and of escaping from their enemies. Such a combination of advantages is not to be found elsewhere among invertebrate animals.

(b) *Adaptability.* No other single class of animals has so thoroughly invaded and colonized the globe as the Insecta. Their distribution ranges from the poles to the equator; every species of flowering plant provides food for one or more kinds of insect, while decomposing organic materials attract and support many thousands of different species. Very many are parasites on or within the bodies of other insects or of some very different animals, including vertebrates. The soil and fresh waters support their own extensive insect fauna. Great heat and cold are not impassable barriers since some species can withstand temperatures of about $-50°$ C, while others live in hot springs at over $40°$ C or in deserts where the midday surface temperature may be twenty degrees higher. A few species live in what seem almost impossible environments; the larva of an Ephydrid fly inhabits pools of crude petroleum in California while some beetles have been reported from argol (containing 80% potassium bitartrate), opium, Cayenne pepper, sal ammoniac and strychnine.

(c) *Size.* The relatively small size of most insects has many advantages. Each individual requires little food so that large populations may occupy small habitats, which often also offer security from enemies. Thus, several leaf-mining larvae may develop in the tissues between the upper and lower epidermal layers of a single small leaf, a weevil will complete its life cycle in one small gorse seed, while a moderate-sized fungus will support very many beetles and fly larvae.

Insects, in fact, vary greatly in size, from minute Hymenopterus parasites about 0·2 mm long to forms like the bulky Goliath Beetle with a length of up to 120 mm. These, however, are extremes and both very large and very small insects suffer disadvantages which do not apply to the more numerous species of an intermediate size. If a very small insect is wetted, the weight and surface tension of the encompassing water film soon exhaust its efforts to free itself. Very large species, on the other hand, are subject to a limitation imposed by their characteristic method of tracheal respiration. Oxygen passes along these breathing tubes by gaseous diffusion and the physical law which this process follows is such that an increase in the size of an insect is not accompanied by a proportional increase in the rate at which oxygen can reach the tissues (p. 84). When an insect reaches a diameter of about 2 cm, therefore, its method of respiration is liable to incommode it and make it sluggish – further increase in bulk would soon make it too inert to survive competition with other organisms. For such reasons the relatively gigantic forms which do occur, as among beetles, grasshoppers, water-bugs and fossil dragonflies, form a very small proportion of their own groups. Furthermore, even among large insects, very few have a diameter of more than about half an inch, though there may be a great extension in the length of the body or the area of flat, plate-like projections from it. Thus the giant stick insect *Pharnacia serratipes* measures up to 260 mm long but retains a proportionately attenuated form. Some of the great fossil dragonflies of Carboniferous times had wings exceeding 2 feet in expanse but with typically slender bodies. The giant Noctuid moth *Erebus agrippina* has a wing spread of 280 mm, but its slender body is no more than 55 mm long, and the same applies to the giant Atlas moths and to Oriental butterflies of the genus *Troides*.

(*d*) *The skeleton.* The skeleton of insects, like that of other arthropods, is an exoskeleton and has many features of great significance. It consists of hard regions or *sclerites* separated by soft membranous zones and therefore combines strength and rigidity with flexibility. It protects the insect mechanically, provides firm sites for the attachment of the muscles which move the body, acts sometimes as a mechanism for storing energy, and is invaginated in various ways to support some internal organs and to line the tracheae and a few other structures.

Its construction in the form of a series of jointed tubes surrounding the body and appendages gives it a much greater power of resistance to bending than the endoskeleton of a vertebrate. The two cases have been contrasted by the Russian writer Chetverikov. In Fig. 1, A represents a cross-section of an insect limb with its tubular exoskeleton while B and C are cross-sections through two vertebrate limbs, each with its axial endoskeleton. It is known from physical principles that the modulus of resistance to bending in a solid cylinder and in a hollow one is given by the two formulae:

$$M = \frac{\pi D}{32} \quad \text{and} \quad M_1 = \frac{\pi(D_1{}^4 - d^4)}{32 D_1}$$

where M and M_1 are the respective moduli and D is the diameter of the endoskeleton, while d and D_1 are the internal and external diameters of the exoskeleton. If, for the sake of argument, we take the case of $d = \frac{4}{5}D_1$ and compare figures A and B, the cross-sectional areas of skeleton and muscle being the same in the two cases, then the formulae show that the limb with the solid axial endoskeleton will be nearly three times weaker than the one with the hollow exoskeleton. Further, it may be shown that to have the same strength as the exoskeleton of Fig. I A, an axial endoskeleton would take up 84% of the total diameter of the limb (Fig. I C); under such conditions there would be little space left for musculature.

Because of its mechanical efficiency, therefore, the insect skeleton combines great strength with lightness. Composed as it is of an amazingly plastic material, it has lent itself to the most varied processes of evolutionary modification under the influence of natural selection. Increased deposition of cuticular substance has occurred in endless ways and in adaptation to manifold requirements. Especially to be noted are the immensely varied

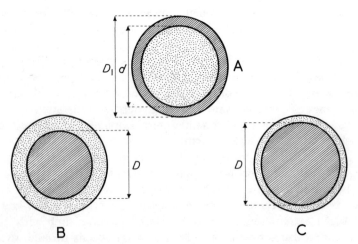

Fig. I. Diagram of cross-sections of limbs; from Chetverikov
A, exoskeleton with internal diameter $d = \frac{4}{5}$ of the external diameter D_1. B and C, endoskeletons. (Skeleton cross-hatched: musculature, etc. stippled)

developments of form and size in the head and jaws; the growth of horns, spines and other processes; of bristles and of scales; of membranous wings and horny elytra; of stout fossorial legs; of needle-like ovipositors, and so forth. It is, furthermore, the exoskeleton which displays many of the structural characters that distinguish the many species of insects.

Because it possesses a rigid exoskeleton, the growing insect must shed it periodically. During this process of moulting the insect is somewhat vulnerable, but this disadvantage may be less real than it seems for the sequence of moults is the basis on which have been evolved the elaborate changes of form which occur in the life cycles of the higher insects and which have allowed the immature and adult stages to become adapted each to its own conditions of life.

(e) *Resistance to Desiccation*. Though some insects are aquatic or inhabit moist environments, the success of the class has depended to a considerable extent on an ability to survive under the relatively dry conditions of terrestrial life. The insects' capacity for resisting desiccation and conserving water shows itself in many ways. The cuticle is provided with a thin layer of waxy material which greatly reduces transpiration from the surface of the body and the openings of the tracheal system, or spiracles, are provided with closing mechanisms or other devices which further reduce water loss, while permitting enough oxygen to enter. The principal excretory products of most insects are insoluble and therefore do not require a large volume of water for their removal from the body in solution, while the terminal part of the alimentary canal reabsorbs water which would otherwise be lost in the faeces. Some insects living under unusually dry conditions depend to a considerable extent on the water produced when foodstuffs are oxidized within their bodies – the so-called metabolic water. Finally, in their reproductive behaviour, with internal fertilization and the habit of laying eggs whose impervious shell protects the developing embryo from desiccation, the insects are well adapted to terrestrial conditions.

(f) *Tracheal Respiration*. The characteristic tracheal respiratory system of insects tends to limit their size and requires modifications to offset this disadvantage and to restrict water loss. In other respects, however, tracheal respiration is very efficient and the direct carriage of gaseous oxygen to within very short distances of the respiring tissues has enabled the insects to evolve the very high rates of metabolic activity needed to achieve rapid flight. Insect flight muscle is, in fact, the most actively respiring animal tissue known and the tracheal system shows several specially interesting adaptations to supply it with enough oxygen (e.g., p. 84). It is also worth emphasizing that though tracheal respiration probably arose and first evolved in terrestrial and aerial habitats, it has nevertheless been retained (though often with a closed tracheal system) in almost all the insects that have returned to colonize aquatic environments secondarily. Only a relatively few, very small insects are able to obtain enough oxygen by simple diffusion through their tissues.

(g) *Complete metamorphosis.* The more highly evolved types of insect life cycle entail a transition from the immature larval stages through a pupal phase into the winged, sexually mature adult (p. 121). This form of development, found for example in beetles, moths and flies, differs from the simpler, incomplete metamorphosis of, say, grasshoppers or cockroaches in that it allows the larva and adult to exploit different food resources and occupy different ecological niches. Such a transition from larva to adult involves extensive anatomical, histological and functional changes, which occur mainly during the apparently quiescent pupal stage. The latter therefore represents an evolutionary innovation of great importance, as can be seen from the fact that insect species with a complete metamorphosis outnumber those with an incomplete metamorphosis by about ten to one and have successfully invaded a much greater range of habitats.

This short preamble reviews the more obvious factors that may have contributed to the success of the insect type of organization. It also helps to explain why that type has persisted from pre-Carboniferous times, with increasing differentiation and expansion, beyond that of any other class of animals. In the pages that follow the elementary features of insect structure and functions are discussed. These are succeeded by a short account of development and metamorphosis, a brief discussion of the more important modes of life, and a section dealing with nomenclature, classification and biology. Some account of the essential features of each of the twenty-nine orders of insects is given, and finally the position of these animals in the arthropod series, their ancestry and their mutual relationships, are dealt with in an elementary way.

2

Anatomy and Physiology

The Integument (Fig. 2). This consists of a cellular layer, the *epidermis*, with an outer non-cellular *cuticle*. The epidermis secretes the greater part of the cuticle and is responsible for dissolving and absorbing most of the old cuticle when the insect moults (p. 119) as well as repairing wounds and differentiating so as to determine the form and surface appearance of the insect. The cuticle forms the outer exoskeleton and is also present as a lining to the fore and hind intestine, to the tracheae, and to other parts similarly formed by an ingrowth of the ectoderm. Typically, it is composed of three layers:

(i) The outermost layer or *epicuticle*, less than 4 μm thick, consists mostly of a hardened protein, but also contains the waxes which are largely responsible for reducing water-loss through the cuticle, as well as an outer 'cement layer'.

(ii) The *exocuticle* is a much thicker layer consisting mainly of chitin and proteins, the latter being 'tanned' by phenolic substances to produce a hard, brown material called *sclerotin* which gives the cuticle its rigidity. The exocuticle is absent or reduced in the more flexible regions of the integument and may be entirely absent from insects with a soft, thin cuticle.

(iii) The *endocuticle*, which is usually the thickest layer, also contains chitin and proteins but the latter are not tanned and this part of the cuticle is therefore soft and flexible.

Both endocuticle and exocuticle consist of numerous laminae arranged more or less parallel to the surface and are traversed by very numerous *pore canals*, each of which may contain a thread-like cytoplasmic extension of the epidermis. Chitin, which makes up 25–60% of the dry weight of the cuticle, is a nitrogenous polysaccharide consisting of many sugar-like residues joined end to end in long molecular chains. It is resistant to alkalis and dilute mineral acids and can be detected by the van Wisselingh test: heating with concentrated potassium hydroxide at 160° C for 20 minutes converts it to chitosan which gives a rose-violet colour with 0·2% iodine in 1% sulphuric acid. In the cuticle the chitin forms ultramicroscopic fibres embedded in a protein matrix.

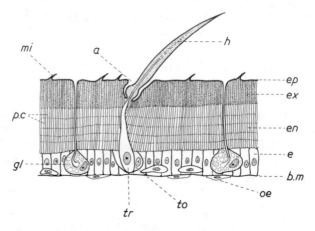

Fig. 2. Integument of an insect, semi-schematic section

a, alveolus; *b.m*, basement membrane; *e*, epidermis; *en*, endocuticle; *ep*, epicuticle; *ex*, exocuticle; *gl*, gland; *h*, hair; *mi*, microtrichia; *oe*, oenocyte; *pc*, pore canals; *to*, tormogen cell; *tr*, trichogen cell

Integumentary Processes. The surface of the cuticle bears two main types of outgrowths:

(i) Rigid non-articulated processes; these include the microtrichia and spines. *Microtrichia* are minute, non-cellular, hair-like structures (Fig. 2), formed entirely of cuticle and often occurring in very large numbers on the wings of certain insects. *Spines* are large, hollow, heavily sclerotized, thorn-like processes of multicellular origin; they are well seen on the legs of cockchafers and dor-beetles (Scarabaeidae).

(ii) Movable articulated processes attached to the cuticle by a ring of articular membrane which may be sunk into a cuticular socket or *alveolus* or elevated on a tubercle; they include macrotrichia and spurs. *Macrotrichia* or *setae* (Fig. 2) are hollow extensions of the exocuticle and epicuticle. Each is secreted by the cytoplasmic outgrowth of a single modified epidermal cell, the *trichogen cell*, while the socket from which the seta protrudes is produced by another specialized epidermal cell, the *tormogen cell*. The following specially modified setae are known: (*a*) *Clothing hairs* which cover the general surface of the body and appendages. They may be branched or *plumose*, as in the bees, or, when specially stiff, they form the bristles of, say, Tachinid flies. (*b*) *Scales*, such as occur in Lepidoptera and some Collembola, Diptera and Coleoptera. Essentially these are flattened setae, often with a striate surface and sometimes containing pigment. (*c*) *Glandular setae*, which serve as outlets for the secretions of epidermal glands; they include the silk-spinning hairs of Embioptera and the urticating hairs of some caterpillars, e.g., those of the Gold-tail moth (*Euproctis chrysorrhoea*). (*d*) *Sensory setae*, which are more or less specialized in structure and have one or more nerve cells at their base. They perceive various stimuli and are discussed further on p. 57. *Spurs* differ from setae in being thick-walled multicellular structures; they are often large and occur more especially on the tibiae of the legs.

Epidermal Glands. These comprise one or more cells specially modified for the secretion of such materials as wax, lac and a variety of substances known as *pheromones* (p. 62) which influence the behaviour or development of other members of the species.

Apodemes. These are sclerotized cuticular ingrowths which collectively form the *endoskeleton*, providing sites for the attachment of muscles and sometimes supporting other organs. They may be more or less tubular or flattened and though the mouths of the invaginations sometimes persist, the apodeme usually becomes solid as the cuticle is laid down. The endoskeleton of the head is known as the *tentorium* (Fig. 3). It consists of paired anterior and posterior arms whose origins are visible externally as slit-like pits; the inner ends of the arms amalgamate to form the body of the tentorium, from near which a third, dorsal pair of arms often arises. The tentorium gives rigidity to the head capsule, provides attachments for muscles and supports the brain and oesophagus. The endoskeleton of the thorax (Figs. 14, 15) usually consists of dorsal phragmata, pleural apodemes and ventral apophyses. The abdominal segments and external genitalia may also bear apodemes.

Coloration. Insect colours fall into three groups: (i) those due to specific pigments (chemical colours), (ii) those due to optical effects produced by special cuticular structures (physical colours) and (iii) those produced by the presence together of pigment and colour-producing structures (combination or physico-chemical colours).

Pigments may be present in the cuticle, the epidermis or the subepidermal tissues and are substances, often of known chemical composition, which are coloured because they absorb some wavelengths of incident light and reflect others. The three commonest groups of insect pigments are the melanins, the carotenoids and the pterines. The black or brown *melanins* of the cuticle are probably derived initially from the amino acid tyrosine by reactions in which the enzyme tyrosinase plays a part; the tyrosinase is generally distributed and melanic patterns depend on the localized occurrence of the tyrosine. The yellow or red *carotenoids* are characteristically synthesized by the plant on which the insect feeds; after ingestion, however, they persist with little or no alteration in the insect's tissues, either free or combined with protein. The *pterines* are a group of white, yellow, orange or red pigments chemically related to the purines; they are found in Pierine butterflies, wasps and other species.

Structural colours differ from pigmentary ones in several respects. They are altered or destroyed by physical changes in the cuticle, they disappear on immersion of the part concerned in liquids of the same refractive index as cuticle (about 1·55), they can be imitated by purely physical models and they cannot be bleached. Scattering, reflection and refraction of the light by particles which are large in comparison with the light wavelengths produce *structural whites*. Optical interference between reflections from a series of superimposed laminae of microscopic thickness produces the characteristic metallic or iridescent *interference colours*, such as those occurring in the wing scales of *Morpho* butterflies or in diamond beetles. *Diffraction colours*, caused by the presence of closely spaced cuticular striae, are characteristic of some beetles.

Combination colours are much more common than purely structural ones. A structural blue may be combined with a yellow pigment to produce a brilliant green, as in some butterflies (*Troides*); or a pigmentary red in the walls of the wing scales of another butterfly, *Teracolus*, combines with a structural violet to give magenta.

SEGMENTATION AND BODY REGIONS

Segmentation. Though the cuticle forms a continuous investment over the whole body of an insect, it usually remains membranous and flexible along certain transverse infoldings, so that the body is divided externally into a series of *segments* separated by *intersegmental membranes*. This segmentation, which is a development of the even more complete metameric segmentation of the early embryo, also manifests itself in some internal organs. Thus the body musculature, nervous system, tracheal system and heart all show, to varying degrees, a longitudinal repetition of parts. The cuticle of a segment or other region of the body may be further subdivided by *sutures*, a general term given to narrow membranous lines of flexibility, or to inflected strengthening ridges (*sulci*), or to impressed lines of no obvious mechanical significance.

The Body Regions. The insect body comprises twenty primitive segments, all of which may be apparent in the embryo, grouped into three well-defined regions or *tagmata* – the head, thorax and abdomen. The *head* is formed of six segments and an anterior non-segmental acron, all closely amalgamated to form a hard case or *head capsule*. It reveals few indications of its segmental origin apart from the possession of paired appendages. The *thorax* consists of three segments; each of these carries a pair of legs and the second and third usually also carry the two pairs of wings. The thorax is connected with the head by the *cervix* or neck, which is largely intersegmental in origin. The *abdomen* comprises eleven segments and a terminal non-segmental *telson*, but reduction and fusion in the posterior region often results in only ten or fewer divisions being visible.

The Divisions of a Segment. In many soft-bodied larvae, such as those of blow-flies and other Diptera, the cuticle is membranous and each segment is a simple ring without division into separate areas. In the majority of insects, however, a typical segment is divisible into four main regions, viz. a dorsal region or *tergum*, a ventral region or *sternum*, and, on either side, a lateral region or *pleuron*. The cuticle of each of these regions may be differentiated into separate sclerites, in which case those composing the tergum are known as *tergites*, those of the sternum are termed *sternites* and those of the pleura are the *pleurites*.

The Appendages. In the embryo each segment typically bears a pair of outgrowths or appendages which may be retained in post-embryonic life. An appendage is a jointed tube implanted in the pleuron of its side. Between two adjacent segments the cuticle is flexible and forms the articular membrane. On account of its jointed structure the whole or part of an appendage is movable by means of muscles. A typical insect appendage consists of a limb-base and a shaft. It is characteristic of the arthropod line to which the insects belong that their appendages are morphologically one-branched or *uniramous* structures; nothing comparable to the biramous appendages of the Crustacea occurs in the insects.

THE HEAD AND ITS APPENDAGES

The Head-capsule. The exterior of the head is formed of several sclerites amalgamated to form a hard compact *head capsule* (Fig. 3). The dorsal region or *epicranium* is commonly divided by a bifurcating suture shaped like an inverted Y. This is the *ecdysial cleavage line*, sometimes known as the epicranial suture, its median part being the *coronal suture* and the two arms the *frontal sutures*. It is a line of thin cuticle along which the head-capsule of the immature insect breaks open at moulting. In the adult stage this is retained only in the more generalized insects; in specialized forms it is absent or replaced by an inflected strengthening ridge. The part of the head above the frontal sutures is the *vertex*; below them lies the *frons*, though the precise limits of the latter area are not easily defined. The frons usually bears the median ocellus and is bounded distally by a transverse *epistomal sulcus* (sometimes absent) which runs between the anterior tentorial pits. Immediately anterior to the frons is the *clypeus*, to which the *labrum* or upper lip is hinged along the *clypeo-labral sulcus*. At the back of the head between the vertex and neck lies the *occiput* while the side walls of the head, below and behind the eyes, are known as the *genae*; the latter may be separated from the facial part of the head by the *frontogenal* and *clypeogenal sulci*. At each side of the clypeus, where it adjoins the genae, is a facet with which the anterior ginglymus of the mandible articulates. Posteriorly each gena bears a cavity for the articulation of the mandibular condyle. The hind surface of the head is perforated by the *occipital foramen*, through which the nerve cord and the oesophagus enter the thorax. Separating

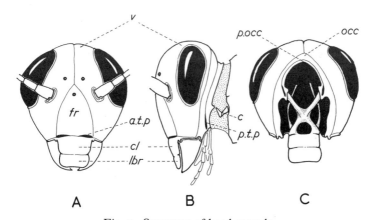

Fig. 3. Structure of head capsule

A, anterior view; B, lateral view; C, posterior view, showing tentorium. *a.t.p.*, anterior tentorial pit; *c*, cervical sclerite; *cl*, clypeus; *fr*, frons; *lbr*, labrum; *occ*, occiput; *pocc*, post-occiput; *p.t.p.*, posterior tenorial pit; *v*, vertex

the labium from the occipital foramen and lying between the genae, there is present in some insects a median sclerite or *gula* (Fig. 79B). Two main types of head occur, differing in the inclination of the long axis and in the position of the mouthparts. In the *hypognathous type*, well seen in cockroaches, grasshoppers and flies, the long axis is more or less vertical and the mouthparts ventral. In the *prognathous type*, prevalent in many beetles, the long axis is approximately horizontal and the mouthparts placed anteriorly.

Head Appendages. These paired structures comprise the antennae, derived from the second embryonic head segment, and the mouthparts, which represent the appendages of the fourth, fifth and sixth segments.

The *antennae* are freely mobile segmented appendages articulated with the head in front of or between the eyes. They differ greatly in appearance in different groups of insects and are sometimes sexually dimorphic, being deeply pectinate in the males of certain moths and plumose in male mosquitoes and midges. The antennae are moved by extrinsic muscles usually arising from the tentorium and inserted on the base of the enlarged first segment or *scape*. Intrinsic muscles arising in the scape are inserted on the *pedicel* or second segment. The remaining divisions of the antenna together constitute the *flagellum* and are entirely without muscles except in the primitively wingless orders Diplura and Collembola. The antennae are sensory organs, well provided with olfactory and tactile receptors (pp. 57, 61).

The mouthparts (Fig. 4) consist of the anterior jaws or *mandibles* followed by a pair of *maxillae* and a lower lip or *labium*, the latter formed by medial fusion of paired maxilla-like structures. Arising from the floor of the mouth cavity is a median, non-appendicular, tongue-like lobe or *hypopharynx*, whose body or *lingua* is supported by suspensory sclerites and, in primitive insects, bears a pair of small outgrowths, the *superlinguae*. The labrum is also closely associated with the mouthparts; its inner surface often bears gustatory receptors and is produced into a stylet-like *epipharynx* in the fleas. The mouthparts show great variation in structure, correlated with different kinds of food and feeding habits. Mandibulate mouthparts, used for biting and chewing, occur in most insects but in the Hemiptera, Siphunculata and some Diptera they are greatly modified for piercing the tissues of plants or animals and sucking up the contained fluids. In most Lepidoptera and some Hymenoptera the mouthparts are long haustellate structures which do not pierce but suck up liquids such as the nectar of flowers. Adult mayflies and some moths do not feed and have atrophied mouthparts.

The *mandibles* are stout and tooth-like in chewing insects, articulating with the head capsule at two points as described on p. 21. Each is moved by abductor and adductor muscles arising from the wall of the head. In the Hemiptera they are produced into long piercing stylets (p. 168) and in the Lepidoptera and most

Diptera they are usually vestigial or absent. Each *maxilla* consists basally of a *cardo* or hinge, which articulates with the head behind the mandible, and a more distal *stipes*. The latter bears two median endites – an outer *galea* and an inner *lacinia* – and a lateral *maxillary palp* composed of several segments. The palp may be borne on a specially differentiated part of the stipes known as the *palpifer*. Highly specialized maxillae occur in the Hemiptera, Lepidoptera, Diptera and some other orders.

The *labium* (fused second maxillae) shows clear indications of its paired origin. The basal *postmentum* (corresponding to the fused maxillary cardines) is often secondarily subdivided into *submentum* and *mentum* (Fig. 4). Distally lies the free *prementum* which is often bilobed and represents the fused stipites. The prementum carries a pair of lateral segmented palps, sometimes arising from special

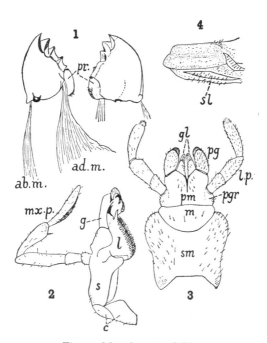

Fig. 4. Mouthparts of *Blatta*

1, Mandibles: *ab.m, ad.m,* abductor and adductor muscles; *pr,* prostheca. 2, Maxilla: *c,* cardo; *g,* galea; *l,* lacinina; *mx.p,* maxillary palp; *s,* stipes. 3, Labium: *gl,* glossa; *l.p,* labial palp; *m,* mentum; *pg,* paraglossa; *pgr,* palpiger; *pm,* prementum; *sm,* submentum. 4, Hypopharynx: *sl,* suspensory selerite

lobes known as *palpigers*, and a median *ligula*. In insects with more generalized mouthparts the ligula consists of paired outer *paraglossae* and inner *glossae*, corresponding respectively to the galeae and laciniae of the maxillae. The *hypopharynx* lies on the inner face of the labium; the common salivary duct opens near its base and in some Diptera it is produced into a long stylet-like organ. Superlinguae occur only in some Apterygotes and in mayfly nymphs.

The **Cervix** or neck is the flexible region between head and thorax; it is largely intersegmental in origin but may include parts of the labial segment and the prothorax. Paired plates, the *cervical sclerites* (Fig. 14), are usually present in the membrane of the cervix. The most important of these are the lateral cervical sclerites which act as a fulcrum between the head and the prothorax; distally they articulate with the occipital condyles of the head and proximally with the episterna of the prothorax. Muscles arising from the head and pronotum are inserted on the lateral sclerites to form a protractor mechanism of the head, the angle between the two lateral sclerites of each side being narrowed or widened, as the case may be.

THE THORAX, LEGS AND WINGS

The three thoracic segments are known respectively as the *prothorax*, *mesothorax* and *metathorax*, and are seen in their simplest form in the Apterygota and many larvae, where they differ little in size or proportions. The acquisition of wings has led to greater specialization and the meso- and metathorax then become more elaborate and more intimately associated. The prothorax is best developed in such insects as the cockroaches and beetles, where its tergum forms a large shield; in higher insects it is often reduced to a narrow ring. The degree of development of the other segments depends on the condition of the wings. Where the fore and hind wings are very similar (as in termites or dragonflies) the meso- and metathorax are about equally developed; conversely, in the Diptera, where only the fore wings are used in flight, the mesothorax is much larger. In referring to the sclerites of the thorax, the segments to which they belong are denoted by the prefixes *pro-*, *meso-* and *meta-*; thus, the protergum refers to the tergum of the prothorax and the mesepimeron to the epimeron of the mesothorax.

In many larvae and pupae and in the Apterygota, the tergum of each segment is a simple, undivided, segmental plate or *notum*. This condition is retained in the prothorax of almost all adult Pterygote insects, but the meso- and metanota of winged forms are usually divided into three sclerites, the *prescutum*, *scutum* and *scutellum* (Fig. 5 A). An additional intersegmental sclerite, the *postnotum*, is commonly also present and bears a phragma to which the dorsal longitudinal flight muscles are attached.

The pleuron possibly originated by modification of a primitive basal leg-segment or *subcoxa*, indications of which are seen in some Apterygota and a few immature stages (Fig. 5 B). In higher insects this region has become enlarged, flattened and fully incorporated into the thoracic wall, to which it gives rigid support (Fig. 5 A). It is then divided into two main pleural sclerites, an anterior *episternum* and a posterior *epimeron*. These are separated by the *pleural suture* or *sulcus*, marking the line along which an internal strengthening pleural ridge is inflected. In a wing-bearing segment the pleuron develops dorsal and ventral articular processes for the wing and leg respectively. Both episternum and epimeron may be subdivided into upper and lower sclerites and a further small pleurite, the *trochantin*, is often present near the lower margin of the episternum.

The sternum (Fig. 5 C) presents many modifications. The main segmental plate is often subdivided into *presternum*, *basisternum* and *sternellum*, the latter two separated from each other by a suture which joins the apophyseal pits, from which the endoskeletal invaginations arise. An intersegmental sclerite, the *spinasternum* or *poststernellum*, may also be visible, bearing a median internal *spina*. In higher insects the sternum is often less fully developed or it may be fused with the pleura so that its boundaries are difficult to define. It is often greatly reduced in breadth and partly infolded between the legs, while the two sternal apophyses are often united on a common base to form a Y-shaped *furca*.

The Legs. The normal insect leg (Fig. 5 B) consists of five segments, the coxa, trochanter, femur, tibia and tarsus and it ends distally in the pretarsus. An additional basal segment, the subcoxa, may – as indicated above – have given rise to the pleuron, but it

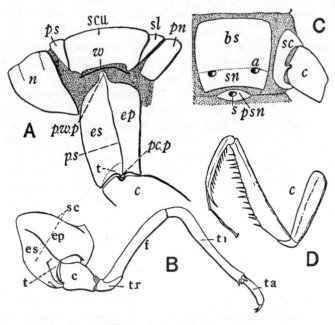

Fig. 5. Thorax and legs

A, Schematic figure showing pronotum *n*, mesonotum and mesopleuron. B, Hind leg and subcoxa of a nymphal Cicada (*adapted from* Snodgrass). C, Schematic figure of a sternum of a wing-bearing segment. D, Fore leg of a Mantid. *a*, pit leading into sternal apophysis; *bs*, basisternum; *c*, coxa; *ep*, epimeron; *es*, episternum; *f*, femur; *p.c.p*, pleural coxal process; *pn*, postnotum; *ps*, prescutum; *p.s*, pleural suture; *psn*, poststernellum; *p.w.p*, pleural wing process; *s*, pit leading into spina; *sc*, hypothetical subcoxa; *scu*, scutum; *sl*, scutellum; *sn*, sternellum; *t*, trochantin; *ta*, tarsus; *ti*, tibia; *w*, wing-base

is not recognizable as a separate, definitive leg segment. The *coxa* is the functional limb-base, articulating with the coxal process of the pleuron and sometimes also with the trochantin and sternum. The *trochanter* articulates with the coxa but its attachment to the femur is fixed; it occasionally appears to be subdivided (e.g., Odonata, many parasitic Hymenoptera; Fig. 91) but the distal piece really belongs to the femur. The *femur* is usually the largest segment and in the hind leg of most Orthoptera it is still further increased in size to accommodate the tibial levator muscles used in jumping. The *tibia* is generally a slender shaft and the tarsus is often divided into 2–5 subsegments or *tarsomeres*, though many larvae and some Apterygota retain the primitive undivided condition. The tarsus often bears ventral pad-like structures, the *plantulae* or tarsal pulvilli. The insect leg ends in the *pretarsus*; in the Protura, Collembola and many larvae this is a simple claw, but in most insects there are paired claws (Fig. 6). The other pretarsal structures, not all of which are usually present together, include the internal *unguitractor plate* on which the flexor (retractor) muscle of the claws is inserted, the paired *pulvilli* which lie one under each claw, a median pad-like *arolium* and a median bristle or *empodium*.

When used for walking, the legs conform to the description given above, but they are often modified structurally to perform other functions. Enlargement of the femur and special modifications of the musculature enable the hind legs to be used for jumping (e.g., grasshoppers, leaf-hoppers and fleas) while aquatic insects have swimming legs which are broad or fringed with specially long hairs. The praying mantids and some other predacious insects have raptorial fore legs, with elongate coxae, spinose femora and tibiae, and reduced tarsi (Fig. 5 D), while some soil-inhabiting forms like the mole-crickets (Gryllotalpidae) have stout, spurred fossorial (digging) legs. In many male insects the legs are modified to hold the female when mating, e.g., the elongate fore legs of male mayflies and the apparatus of suckers on the fore tarsus of some male Dytiscidae.

Walking. When they walk or run, insects move the legs of each side in a stepping sequence which is basically similar in many species: the hind, mid and fore legs are moved forward one after the other. However, the phase difference between the stepping sequences of the right and left sides of the body differs from one case to another, so that a great variety of gaits can result. Some legs may step in pairs or there may be three legs in motion at any one time – two on one side and one on the other. It sometimes happens in this way that the insect is momentarily supported on a tripod formed of the fore and hind leg of one side and the mid leg of the other while the three remaining legs are in motion. This alternation of 'tripods of support' is not, however, a constant feature of insect locomotion and other patterns of support occur, depending on the time-relations of the individual leg movements. The complicated set of coordinated leg movements seems to be controlled by a fundamental, overall pattern of activity in the

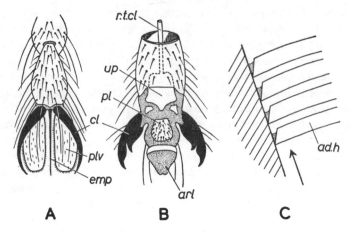

Fig. 6. Pretarsus and adhesive organs

A. Pretarsus of *Asilus crabroniformis*. B. Pretarsus of *Apis mellifera* (*after* Snodgrass). C. Apices of hairs from adhesive organ of *Rhodnius prolixus* (*after* Wigglesworth); sliding movement up a steep surface in the direction of the arrow takes place readily; movement in the opposite direction is greatly hindered by surface forces depending on the presence of a film of oily secretion beneath the obliquely truncate tip of each hair. *ad.h*, adhesive hair; *arl*, arolium; *cl*, claw; *emp*, empodium; *pl*, planta; *plv*, pulvillus; *r.t.cl*, retractor tendon of claws; *up*, unguitractor plate

central nervous system, monitored and modified as necessary by local reflexes involving each separate leg.

Many insects are able to climb and adhere to steep smooth surfaces by the presence of adhesive organs on the legs. These are usually the plantulae of the tarsus and the arolium and pulvilli, but *Rhodnius* and its relatives have a special adhesive organ at the apex of the tibia. Though the physical basis of adhesion is uncertain, it may be due to surface molecular forces which operate when the extremities of the fine hairs clothing the adhesive organs are moistened by a glandular secretion and brought into contact with the substrate (Fig. 6).

Wings. Most adult insects have two pairs of wings, articulated by a complex group of sclerites to the two sides of the meso- and metathoracic terga respectively. The primitive Apterygota, however, are all wingless, representing an evolutionary stage which precedes the origin of these structures, while some Pterygote insects, such as fleas and lice or worker ants and termites, have lost their wings in the course of evolution from winged ancestors. The youngest developmental stages of insects have no wings but their rudiments become apparent in later instars, two main types of wing development being recognizable. In the Exopterygotan insects the wings arise as externally visible, projecting wing pads which gradually increase in size at each moult, whereas in the Endopterygota they undergo their early development concealed beneath the cuticle of the larva and first become visible externally in the pupa. Histologically, however, the two types of developing wing rudiments are very similar. They are flattened evaginations of the lateral margins of the meso- and metaterga, each of whose main surfaces consists of epidermis and basement membrane, the former secreting the cuticle which invests the wing once it comes to project externally. During development the upper and lower epidermal layers meet and fuse except along certain linear channels or *lacunae*. Tracheae and nerves grow into the lacunae, blood circulates in them and they determine the course of the strengthening tubes or *wing veins* which eventually appear in the fully formed wing. At the moult into the adult the relatively small, fleshy wing rudiments are inflated to their full size by blood pressure and the epidermal cells atrophy so that the adult wing comprises a thin double layer of cuticle supported by the more heavily sclerotized wing veins. The wing surfaces may be smooth or clothed with microtrichia (Fig. 2) or macrotrichia; or they may be partially scaled (mosquitoes) or wholly scaled as in Lepidoptera. A small scale-like sclerite or *tegula* (Fig. 91) overlaps the base of the fore wing in Lepidoptera, Hymenoptera and some Hemiptera.

In the more generalized insects the two pairs of wings are used in flight and are very similar in size and shape, though the posterior part of the hind wing is often enlarged to form a fan-like expansion or *anal lobe* which is delimited from the more anterior part of the wing by an *anal furrow*. In other insects, one pair of wings – usually the fore wings – is specialized, often to protect the delicate hind wings. Thus in the cockroaches, mantids, stick-insects and Orthoptera the fore wings are sclerotized, leathery *tegmina*; in the Heteroptera the fore wings are sclerotized over their basal half and are known as *hemelytra*, while in the earwigs and beetles they form hard sclerotized protective structures known as *elytra* and are no longer used for flight. In the Diptera the hind wings are modified to form small vibrating organs known as *halteres*, which control equilibrium during flight and, in the more specialized flies, are largely concealed by the lobe-like *calypters* at the base of the fore wings. In many insects the fore and hind wings of each side are held together in flight by some form of *coupling apparatus* so that they beat in unison. In many moths this apparatus consists of one or more stout bristles, the *frenulum*, which arises from the base of the hind wing and interlocks with a *retinaculum* on the under side of the fore wing (Fig. 88). In the Hymenoptera a series of minute hooks or *hamuli* on the front margin of the hind wing engages with the reflexed hind border of the fore wing and in this way unites the two wings of a side (Fig. 91).

Wing venation. The complete system of wing veins is termed the *venation*. It is virtually constant for each species and the various taxonomic groups often show characteristic venational features which are therefore of great importance in insect classi-fication. Though many specialized kinds of venation are known, it has been possible to formulate a general system (Fig. 7) from which all others appear to have been derived. The homologies of veins within this scheme may frequently be decided on three main types of evidence: (*a*) That provided by the system of tracheae found in the pupal or nymphal wing pad. These often agree broadly in arrangement with the subsequent venation but retain other, more primitive, features which make them easier to identify than are the veins. (*b*) That provided by the venation of fossil insects or by comparative studies of the more primitive recent groups, thus revealing the connexions between primitive, easily identifiable, patterns and the more specialized kinds. (*c*) That due to the fact that in the more generalized fossil and recent insects the wings tend to be longitudinally folded, or plicate, with inter-vening furrows, e.g., Ephemeroptera. Veins found on the ridges – when the outspread wing is viewed from the dorsal side – are termed *convex veins* (indicated +) and those in the furrows are *concave veins* (indicated −). The relations of these alternately con-vex and concave veins are shown in Fig. 7.

It should be noted that none of the above three criteria are wholly satisfactory and in some groups (e.g., Hymenoptera) the homologies of the wing veins are not established with certainty.

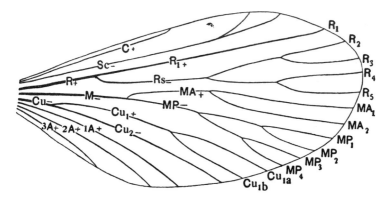

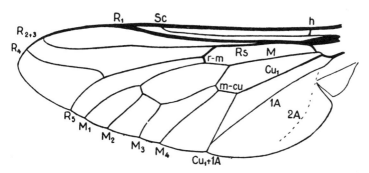

Fig. 7. *Above*. Hypothetical primitive type of venation (convex veins +
and concave veins −). *Below*. Wing of horse-fly (*Tabanus*)
For reference lettering, see pp. 31–2

Nevertheless, the system of terminology depicted for a hypothetical primitive system of wing veins in Fig. 7 is now widely
accepted. It has largely replaced the older special systems of
nomenclature, each applicable only to a particular group, though
the latter still retain the advantage of convenience for some
descriptive taxonomic purposes. Beginning from the anterior or
costal margin of the wing, the first vein is the *costa* (C), which
is unbranched and forms the wing margin. The *subcosta* (Sc) lies
closely behind the costa and is usually undivided. The *radius* (R)
or third vein forks into an anterior branch R_1 and a posterior

branch or *radial sector* (Rs), which divides into four branches, R_2 to R_5. The fourth vein or *media* (M) forks into an *anterior media* (MA) which is typically 2-branched, and a *posterior media* (MP) which is 4-branched. The fifth vein or *cubitus* (Cu) likewise divides into anterior and posterior branches. The *anterior cubitus* (Cu_1) is typically 2-branched and the *posterior cubitus* (Cu_2) is undivided. Finally, a variable number of *anal veins* (1A, 2A, etc.) are present. Additional rigidity of the wing membrane is obtained by the development of a network of *cross-veins* between the veins. While these are numerous and variable in the Ephemeroptera and Odonata, in the higher orders they tend to become few and located at fixed positions of mechanical advantage. The main cross-veins and their symbols are the *radial* (r) from R_1 to Rs; the *radio-medial* (r-m) from R to M; the *medial* (m) from MP_2 to MP_3; and the *medio-cubital* (m-cu) from MP to Cu. The veins divide up the wing area into *cells* and the name of each cell is taken from that of the vein which forms its anterior border. Thus, the cell lying behind the main stem of the media is cell M, while a cell lying behind R_1 is cell R_1. Where two veins have fused the cell immediately behind is named from the posterior component; thus, when veins R_4 and R_5 coalesce the area behind is cell R_5.

The hypothetical venational pattern is most nearly approached in some Palaeozoic fossils. Among living insects the least departure from it is seen in the Ephemeroptera and some primitive Endopterygotes like the Trichoptera and Mecoptera. In most recent insects the media is represented only by MP, but the Ephemeroptera retain the primitive condition while in the Odonata and perhaps also the Plecoptera it is only MA which has been retained. In some orders, e.g., Neuroptera, specialization has taken place by the addition of subsidiary branches to the main veins and in the Ephemeroptera there are many secondary *intercalary veins* lying between the main ones. More often, however, specialization occurs by reduction in the number of main veins and their branches and in some parasitic Hymenoptera, for example, the venation has atrophied entirely. A small, thickened, darkly pigmented area near the costal margin of the fore wing in many Hymenoptera and on both pairs of wings in Odonata is known as the *stigma* or *pterostigma* (Fig. 66).

Origin of Wings. The *tracheal gill* theory claims that wings were derived from plate-like thoracic gills of the kind well shown on the abdomen of some mayfly nymphs. Being basally articulated with the body and already supplied with muscles and tracheae, it is claimed that they only required to become enlarged and adapted for flight. Wings, however, differ in their mode of development and do not seem to be serially homologous with tracheal gills. The theory also involves the assumption that the ancestors of winged insects were aquatic, which is contrary to much evidence. The more widely accepted *paranotal theory* postulates that wings arose from lateral tergal expansions, or *paranota*, of the thorax. It is maintained that the prothoracic lobes of some of the most ancient fossil insects are organs of this kind which had persisted long after the paranota of the other thoracic segments had developed into wings. Paranotal expansions occur in positions characteristic of wings, not only on the thorax but also on the abdomen in various arthropods; among insects they are seen in many larvae and also in *Lepisma*, where they contain tracheae recalling those of a wing pad. It is suggested that they became sufficiently large to function as gliding planes in leaping insects. Later they acquired basal articulations which, along with the development of muscles, enabled them to become organs of independent movement.

Flight. Insects fly at various speeds. These are often exaggerated and probably no insects surpass the larger dragonflies which may move at about 10 metres per second. The movements of the wings during flight are quite complex. The trajectory of the wing tip relative to the insect's body takes the form of an elongate figure-of-eight more or less oblique to the long axis of the body. During the downstroke – on which most of the propulsive force is generated – the wing is pulled downward and forward and its surface assumes a position in which the anterior margin is lowered in relation to the elevated posterior area. In the upstroke the wing is pulled upward and backward while its anterior margin now comes to lie relatively higher than the depressed posterior area (Fig. 8). The result of these movements is that aerodynamic forces act on the wings in such a way as to maintain the insect in the air and propel it forwards. The movements of the wings are brought about through the activity of three sets of flight muscles: the direct, indirect and accessory indirect muscles. The *direct* flight muscles arise on the pleural and sternal regions and act directly on the base of the wing since they are inserted on the basalar and subalar sclerites or on the axillary sclerites. They are important in the Orthoptera, Dictyoptera, Odonata and Coleoptera but elsewhere they are small and the main propulsive forces are developed by the powerful *indirect* flight muscles. These are arranged in two functional groups – a dorsal longitudinal set running between the phragmata of meso- and metathorax and a dorsoventral set running from tergum to sternum. The traditional explanation of the mode of action of these muscles is illustrated in Fig. 8 A and B; contraction of the dorsoventral muscles lowers the tergum and elevates the wings (Fig. 8 A) while contraction of the dorsal longitudinal muscles raises the tergum and depresses the wings (Fig. 8 B). This is an oversimplified interpretation; the pterothoracic box consists of rigid and more flexible parts and may be deformed in a complicated manner while elastic forces thus generated in the cuticle play an important role in moving the wings. Certain *accessory indirect* muscles, such as the pleurosternals and tergosternals, also influence flight movements by altering the relative position of moving parts or by changing the elastic properties of the pterothorax. In the fly *Sarcophaga* the mechanism of wing movement has been studied in detail and may be understood by reference to Fig. 8 C and D, which show the changing relations between the scutum, the first and second axillary sclerites and the pleural wing process. When the wing is elevated (Fig. 8 C), contraction of the dorsal longitudinals tends to force the edge of the scutum (f) laterally but this force is resisted by the contraction of the pleurosternal muscles and little movement of the wing takes place. Another effect of the contracting dorsal longitudinal muscles, however, is to exert an upwardly directed force on the first axillary via a process of the scutellum (which articulates with the axillary at z). When this force becomes sufficient to carry the point y nearly to the line joining x and f, the elastic energy stored in the tergum is suddenly released through the rotation of the second axillary about the pleural wing process and the wing 'clicks' into the depressed position (Fig. 8 D). The contraction of the dorsoventral muscles then follows and this also has the effect of forcing the point f laterally; again, however, there is no appreciable movement of the wing until the force on z – now acting in a *downward* direction – reaches a critical value, when the wing suddenly clicks back to the elevated position.

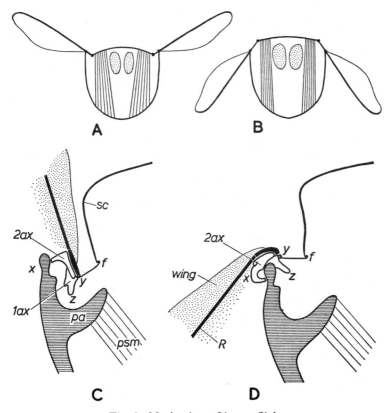

Fig. 8. Mechanism of insect flight

A and B illustrate the traditional theory of the mode of action of the indirect flight muscles. C and D are diagrammatic transverse sections through base of wing of *Sarcophaga*, showing flight movements (*after* Boettiger & Furshpan; see text for explanation). 1 *ax*, 2 *ax*, 1st and 2nd axillary sclerites; *pa*, pleural apophysis; *psm*, pleurosternal muscle; *R*, radial vein; *sc*, scutum of mesothorax; *f*, *x*, *y* and *z* are points referred to in text

The frequency with which the wings beat varies greatly in different insects. In a dragonfly there may be only 28 strokes per second and no more than 9 per second in a cabbage butterfly. On the other hand, the number of strokes per second is stated to be 180–200 in the housefly and the honey bee and 280–310 in a mosquito; a few species even exceed these figures considerably. The necessary high rates of muscular contraction are only possible because of the peculiar physiological properties of some insect flight muscles which behave in an asynchronous manner. Such muscles are capable of contracting as a direct response to being stretched and the frequency with which they undergo cycles of contraction and relaxation is therefore no longer limited by the relatively slow rate at which they receive nervous impulses. During flight the orientation of the insect is preserved or altered by various nervous mechanisms. For example, the halteres of Diptera, which vibrate in flight, are sense organs equipped basally with groups of campaniform sensilla and enable the insect to maintain its equilibrium after the manner of a gyroscopic control mechanism. Prolonged flight is possible only if the insect's body contains adequate food reserves to supply the required metabolic energy. In general, insects with asynchronous flight muscles, such as the Diptera and Hymenoptera, use carbohydrate reserves, while those with synchronous muscles employ lipid reserves. Locusts, aphids, and some others, however, use carbohydrate reserves at first and then turn to lipids, while the tsetse fly *Glossina* relies on oxidation of the amino acid proline.

THE ABDOMEN AND GENITALIA

The insect abdomen consists primitively of eleven segments and a terminal non-segmental *telson*. These twelve divisions are usually recognizable in the embryo but the full complement is rarely seen in the postembryonic stages. Only the Protura retain a distinct telson and this order is also peculiar in showing *anamorphosis* – the newly hatched stage has eight segments and telson but subsequent stages acquire another three segments which develop from the front of the telson. The Collembola also display an unusual condition; they have only six abdominal segments, both in the embryo and afterwards. Eleven segments may be recognized in the Thysanura and most generalized Pterygotes, but in many specialized insects the first one to three segments may be reduced and the tenth and eleventh are inconspicuous and often fused together. It is sometimes convenient to distinguish the genital segments – the eighth and ninth in the female and the ninth in the male – from the pregenital and postgenital ones. When the eleventh segment is well developed it may be divided into a dorsal *epiproct* –

the tergum – and a pair of *paraprocts* – the divided sternum (Fig. 10).

In the embryos of many insects each abdominal segment bears a pair of appendages, a varying number of which are retained in a reduced or highly modified condition in young and adult Apterygotes. Thus, in the Machilidae (Thysanura) many of the sterna bear a pair of plate-like limb-bases or *coxites*, each carrying a *style* very like those on the coxae of the second and third thoracic legs (Fig. 61). In the order Diplura the styles are present but the coxites are indistinguishably fused with the sterna and it is probable that the definitive sterna of higher insects are almost all of this composite type. It is possible that the leg-like abdominal processes of caterpillars and some other immature stages are true segmental appendages, but the adult Pterygote insects only retain recognizable appendages on the genital segments – where they form the external genitalia discussed below – and on the eleventh segment. The appendages of the eleventh segment, the *cerci*, may be long and segmented; they are sometimes inserted between the epiproct and paraprocts (Fig. 10) but when the eleventh segment is atrophied they appear to arise from the tenth. In the Thysanura and most Ephemeroptera the epiproct is prolonged into a median caudal filament flanked by the two similar cerci (Fig. 9). The aquatic nymphs of the Zygopteran dragonflies also have three similarly situated processes which form flattened leaf-like gills (Fig. 67), but it is probable that the two lateral ones are outgrowths of the paraprocts rather than modified cerci. In many higher insects the cerci are absent.

External Genitalia. In both sexes these are usually said to be derived from highly modified abdominal appendages. Primitively, each genital appendage or *gonopod* may be regarded as consisting of a limb-base or *coxite* (gonocoxa) which bears an apical *style* (gonostylus) and has a long process or *gonapophysis* arising from its median basal region. Some or all of these parts may, however, be difficult to recognize in the adult insect, and though studies of the musculature and postembryonic development provide much help, the homologies of the various structures which make up the external genitalia are not known with certainty in all orders of insects.

In the male the genital segment is the ninth abdominal one and its appendages typically form the genitalia. When complete these comprise (*a*) a pair of lateral *claspers* which grip the female in copulation and between which lies (*b*) the median *aedeagus* or *penis* which is flanked by (*c*) a pair of *parameres*. The penis is the intromittent organ of Pterygote insects; it bears the gonopore and develops by fusion of a pair of outgrowths from or behind the ninth segment. Only in mayflies and some primitive earwigs, however, is it obviously paired in the adult (Fig. 9 A). Parameres are present in many orders (e.g., Thysanura, Fig. 61); they are probably not always homologous but in some cases they and the paired rudiments from which the penis develops perhaps represent the divided gonapophyses of the ninth abdominal segment. In the Thysanura the coxites and styles of the ninth abdominal segment resemble those of the pregenital segments and are not used for sperm transfer. In some insects, however, such as the Ephemeroptera (Fig. 9 A), the coxites and styles form functional claspers. Among other groups, where claspers as such are lacking, the coxites fuse indistinguishably with the ninth sternum to form the *subgenital plate* or *hypandrium* and the styles – if they persist – are

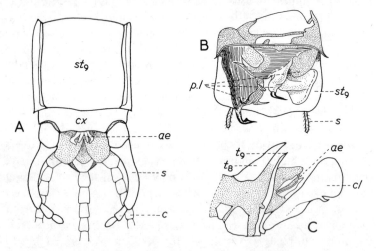

Fig. 9. Structure of male genitalia

A, a mayfly (Ephemeroptera), ventral view; B, *Periplaneta*, dorsal view, after removal of terminal abdominal segments (after Snodgrass); C, a moth (*Cydia pomonella*) (after Snodgrass). *ae*, aedeagus; *c*, cercus; *cl*, clasper; *cx*, coxite of ninth sternum; *p.l*, penis lobes; *s*, style of ninth abdominal sternum; st_9, ninth abdominal sternum; t_8, t_9, eighth and ninth abdominal terga

small simple structures, e.g., some Orthoptera and the cockroaches (Fig. 9 B). In the Lepidoptera the claspers (usually known as valves) probably represent only the styles, the coxites having become amalgamated with the sternum (Fig. 9 C); it has, however, been claimed that the claspers of Lepidoptera, Mecoptera and primitive Diptera are parameres and are not homologous with the claspers of mayflies. Some orders, e.g., the Plecoptera and Coleoptera, have no recognizable traces of claspers though in the former order – and in some other groups – the functional copulatory organs include structures which are not derived from the genital appendages. The male genitalia of related species and genera often differ considerably in detail and are therefore of great taxonomic value.

In the female, the primitive condition of paired gonopores opening behind the seventh abdominal sternum is retained in the Ephemeroptera, but most female insects have only a single genital opening behind the eighth sternum. The paired appendages of the genital segments (eighth and ninth) then typically form the *ovipositor* or egg-laying organ. In the Machilidae (Fig. 61) this consists of the two closely associated pairs of annulate gonapophyses, each arising from the base of the appropriate unmodified coxite. Generally, however, the ovipositor consists of three paired valves associated basally with one or two pairs of small plate-like sclerites often formerly referred to as valvifers (Fig. 10). The *anterior* or *ventral valves* probably represent the gonapophyses of the eighth segment though they are connected basally with a small sclerite (the *gonangulum*) thought to have been derived from part of the ninth coxite. The *posterior* or *inner valves* and the *lateral* or *dorsal* ones are respectively the gonapophyses and parts of the coxites of the ninth segment. The styles are absent and the extension of the ninth coxite that forms the dorsal valve is sometimes called the *gonoplac* by insect morphologists. In some Orthoptera all three pairs of valves are present, often interlocking by tongue-and-groove joints to form a rigid egg-laying organ (Fig. 10); in other Orthoptera the inner valves are rudimentary. In many Hemiptera and Hymenoptera the dorsal valves are separate and ensheath the effective ovipositor (formed from the other two pairs) when this is not in use. In all cases a space between the valves of the opposite sides forms an egg channel down which the eggs pass when being laid.

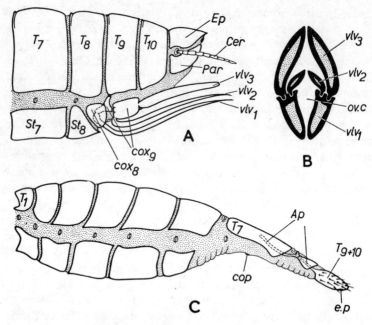

Fig. 10. Ovipositor and abdomen

A. Diagrammatic lateral view of a primitive insect ovipositor. B. Transverse section through oviposior of a Tettigoniid, showing relative positions of valves. C. Abdomen of *Lymantria monacha*, with terminal segments extended (*after* Eidmann). *Ap*, apodemes; *Cer*, cercus; *cop*, copulatory aperture; cox_8, cox_9, coxites of 8th and 9th segments; *Ep*, epiproct; *e.p*, egg pore; *ov.c*, egg channel; *Par*, paraproct; St_7, St_8, 7th and 8th sterna; T_1, etc., terga of 1st, etc., segment; vlv_1, vlv_2, vlv_3, anterior, inner and lateral valves of ovipositor

In higher Hymenoptera such as the wasps and bees the ovipositor is no longer used for laying eggs but has become modified to form a poison-injecting apparatus, the *sting* (Fig. 11). At rest this lies in a pocket within the seventh abdominal segment, from which it is exserted when in use. At the base of the sting are three pairs of plate-like sclerites – the *quadrate*, *oblong* and *triangular plates*. These articulate with each other as shown in Fig. 11 and represent respectively the ninth tergum, the ninth coxite and the gonangulum. The dorsal valves are soft, palp-like organs which arise from the posterior part of the quadrate plate and ensheath the protrusible shaft of the sting in the resting position. The shaft consists of the paired, barbed *lancets* (modified anterior valves) and

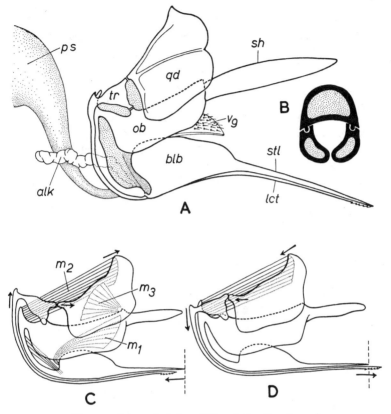

Fig. 11. Sting of *Apis mellifera* (*after* Snodgrass)

A. Sting and associated glands in lateral view. B. Transverse section through stylet and lancets. C, D, Mode of action of sting; arrows indicate directions of movements; for explanation, see text. *alk*, alkaline gland; *blb*, bulb of stylet; *lct*, lancet; m_1, protractor muscle of shaft; m_2, protractor muscle of lancet; m_3, retractor muscle of lancet; *ob*, oblong plate (or ninth coxite); *ps*, poison sac; *qd*, quadrate plate (or ninth tergum); *sh*, sheath; *stl*, stylet; *tr*, triangular plate (or gonangulum); v_9, ventral part of 9th segment

the unpaired *stylet* (fused inner valves). At their bases the lancets are attached to the triangular plates while the stylet is expanded into a bulb-like structure from which the free proximal part of each inner valve runs to the anterior region of the corresponding oblong plate. Lancets and stylet are associated as shown in Fig. 11 B, the lancets sliding to and fro along ridges on the ventral

face of the stylet when the insect stings. The mechanism of stinging is as follows: First the basal apparatus is swung downwards
on a pivot between the postero-dorsal corner of the quadrate plate
and the eighth tergite; meanwhile the shaft is depressed by the
contraction of muscles which run from the oblong plate to the bulb
of the stylet (Fig. 11 C, m_1); finally the contraction of powerful
muscles (Fig. 11 C, m_2) running from the quadrate plate to the
anterior part of the oblong plate causes the rotation of the triangular plate so as to push out the attached lancet. Retraction of the
lancet is brought about through the action of a muscle (Fig. 11 C,
m_2) running from the quadrate plate to the posterior part of the
oblong plate. The muscles on the two sides of the sting work
alternately and by successive acts of protraction and retraction the
lancets are driven more and more deeply into the victim's body,
the stylet following the lancets into the wound. The poison which
is injected into the wound is secreted by a pair of long thread-
like glands in the abdomen. Their secretion accumulates in a large,
ovoidal *poison sac* which opens at the base of the sting into the
poison canal between stylet and lancets (Fig. 11 B). In the honey
bee the poison contains a protein and certain enzymes which cause
the tissues of the victim to produce histamine; the latter gives rise
to the usual symptoms of a bee sting. The function of the so-called
alkaline gland, which also opens at the base of the sting, is uncertain.

Among many insects, e.g., the Ephemeroptera, Plecoptera,
Lepidoptera, Diptera and Coleoptera, there is no true appen-
dicular ovipositor. In some members of the three latter orders,
however, the posterior segments of the abdomen form a slender
telescopic tube or *oviscapt* (Fig. 10) which acts as the functional
ovipositor. It bears the opening of the egg passage distally and
enables the eggs to be laid in crevices or other concealed situations.

THE MUSCULAR SYSTEM

The muscles of insects, both skeletal and visceral, are striated in
all cases, thus differing from those of vertebrates and many other
animals. Even the delicate fibrillae of the heart wall and the
muscle reticulum around the gut and other viscera are seen to be
striated when suitably stained. Muscles are fibrous structures
composed of many protein arrays or *myofilaments*, arranged into

myofibrils. These in turn are grouped into muscle *fibres*, each composed of *fibrils* embedded in a nucleated cytoplasmic matrix, the *sarcoplasm*. This may form a peripheral layer, the *sarcolemma*, or make up a central cylindrical core around which the myofibrils are arranged. Each myofibril consists of alternating light (isotropic) and dark (anisotropic) portions, so giving the whole fibre its cross-striated appearance. The skeletal muscles are attached to the integument in various ways (Fig. 12). Sometimes they are connected with unmodified epidermal cells but more often the latter are traversed at the site of attachment by *tonofibrillae* – intracellular filaments which are closely associated with the myofibrils and sometimes appear more or less continuous with specially differentiated regions of the overlying cuticle. The more powerful skeletal muscles are often attached to special cuticular ingrowths which may take the form of tubular or rod-like apodemes (Fig. 12) or flat plate-like phragmata (Fig. 14).

Some insects are able to raise weights heavier than their own bodies or can leap relatively great distances and this has given rise to the popular idea that their muscles are far more powerful

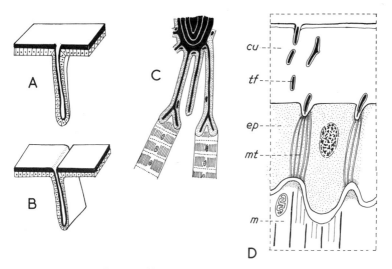

Fig. 12. Skeleto-muscular structures

A, an apodeme; B, a phragma; C, unicellular apodemes with attached muscle fibres (after Snodgrass); D, ultrastructure of muscle attachment to integument (after Caveney). *cu*, cuticle; *ep*, epidermal cell; *m*, muscle; *mt*, microtubules traversing epidermis; *tf*, tonofibrillae running through cuticle in pore canals.

than those of vertebrates. The insects' performance must, however, be considered in relation to the size of their bodies. The power of a muscle varies with its cross-sectional area, i.e., with the square of a linear dimension. But the volume (and hence the mass) of the insect's body varies as the cube of a linear dimension. Consequently, other things being equal, the *relative* muscular power must increase as the size of the animal diminishes. The *absolute* power of a muscle, on the other hand, is the maximum load it can raise per square centimetre cross-section and this tends to be a little less in insects than in vertebrates. In most physiological properties there are no great differences between insect and vertebrate muscles and the biochemical changes accompanying contraction are essentially similar in the two groups though there are some interesting adaptive differences. Insects flying for long periods can sustain intense metabolic activity in their flight muscles without incurring an appreciable 'oxygen debt'. This completely aerobic metabolism is accompanied by an unusually high rate of ATP production and high rates of fuel consumption (p. 36). In insects that can fly actively the flight muscle mitochondria are very numerous, very large, and provided with very densely folded cristae, thus supplying a large surface area for the rapid exchange of metabolites between the oxidative mitochondrial enzyme systems and the general muscle cytoplasm. As mentioned on p. 35 the very rapid wing movements of some insects depend on a special mechanism whereby the flight muscles, which are arranged in antagonistic sets, can contract as a direct response to being stretched. They are thus able to undergo repeated cycles of contraction and relaxation with a frequency far greater than that with which they could be stimulated by motor nerve impulses.

Myology

The following account enumerates the main skeletal muscles of a generalized insect, almost all of which are paired because of the bilateral symmetry of the body. Each muscle is attached at one end – its *origin* – to a more or less stationary part of the exoskeleton and at the other end – its *insertion* – to the part which is moved. So far as the segmented appendages are concerned, the muscles may be *extrinsic* or *intrinsic*. Extrinsic muscles arise outside the limb, are inserted near its base, and are mainly concerned with its movements as a whole. Intrinsic muscles have their origins and insertions within the limb and effect movements of individual segments or parts. The various muscles may be named from their sites of origin and insertion – as when one speaks of a pleurocoxal muscle – or by reference to their functions, e.g., the tibial levator. The latter system is probably the most generally useful, though it may lead to confusion when studying muscle homologies since small differences in position may lead to important changes in function. Muscles whose contraction causes forward and backward movements of a limb in the horizontal plane are known as *promotors* and *remotors* respectively. *Levators* and *depressors* respectively raise or lower an appendage in the vertical plane, e.g., the levator muscles of the antenna raise that appendage. *Flexors* draw one part towards another (e.g., a leg segment towards the adjacent one) while the antagonistic *extensors* move them apart. In some cases, depending on the spatial relations of the part concerned, the terms flexor and extensor are synonyms of depressor and levator. *Adductors* draw an appendage towards its fellow of the opposite side of the body, whereas *abductors* move the two apart. Finally, *rotators* bring about turning movements of a limb or part.

1. **The Head Muscles.** The chief muscles of the head and its appendages are as follows (Fig. 13):

(*a*) The *cibarial* and *pharyngeal dilators* run respectively from the clypeus to the dorsal wall of the cibarium and from the frons to the pharynx. One or both may operate the sucking pump of specialized insects (e.g., Hemiptera, Dytiscid beetle larvae).

(*b*) *Labrum.* The paired *anterior labral* (levator) arises on the frons and is inserted on the anterior face of the labrum. A pair of *posterior labrals* (depressors) has the same origin but inserts on the posterior surface of the labrum; each runs lateral to the corresponding anterior labral.

(*c*) *Antenna.* Inserted on the base of the scape are the extrinsic antennal muscles – a *levator* and *depressor* – both arising from the tentorium. The only other antennal muscles in most insects are the intrinsic muscles of the scape, mainly *levators* and *depressors* of the *flagellum*, which arise within the base of the scape and insert on the base of the pedicel. In the Diplura and Collembola, however, intrinsic muscles are present in each flagellar segment except the last.

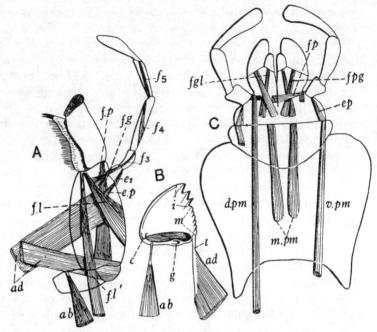

Fig. 13. Muscles of head appendages of an Orthopteroid insect

A, maxilla, right: dorsal (anterior) view. B, Mandible, left. C, labium: dorsal (anterior) view. *ab*, rotator or abductor; *ad*, adductor; *c*, condyle; *d.pm*, dorsal premental muscle; e_2, extensor of 2nd palpal segment; *ep*, extensor of palp; $f_3 f_4 f_5$, flexors of 3rd, 4th and 5th palpal segments; *f.l*, stipital flexor of lacinia; *f.l'*, cranial flexor of lacinia; *f.g*, flexor of galea; *f.gl*, flexor of glossa; *f.p*, flexor of palp; *f.pg*, flexor of paraglossa; *g*, ginglymus; *i*, incisor area; *m*, molar area; *m.pm*, median premental muscles; *t*, tendon; *v.pm*, ventral premental muscle

(d) *Mandible.* A powerful *adductor* arises dorsolaterally on the wall of the head and is inserted by a strong apodeme on the inner basal margin of the mandible. A smaller *abductor* arises external to the adductor and is inserted on the outer basal angle of the mandible.

(e) *Maxilla.* The *rotators of the cardo* arise from the dorsal wall of the head and are inserted on the cardo while *adductor* muscles inserted on the cardo and stipes originate from the tentorium. The *cranial flexor of the lacinia* is a long slender muscle arising from the occipital region of the head. On the stipes there originate the *stipital flexor of the lacinia*, the *flexor of the galea* and the *levator* and *depressor of the palp*. The individual palp segments may contain *intrinsic palp muscles*.

(*f*) *Labium*. The extrinsic muscles of the labium are the *dorsal* and *ventral premental muscles*; these arise from the tentorium and are inserted respectively on the distal and proximal boundaries of the prementum. A pair of *median premental muscles* arises on the postmentum or submentum and is inserted on the proximal margin of the prementum. The glossae, paraglossae and palps are moved by muscles arising in the prementum and corresponding to the stipital muscles of the maxilla. The cranial flexor of the lacinia has no counterpart in the labium.

2. **The Thoracic Muscles.** These muscles (Figs. 14 and 15) include:

(*a*) *Dorsal longitudinal*, attached to successive phragmata; they are best developed in the wing-bearing segments, where they form the main depressors of the wings. In apterous insects and in those with weak flight these muscles are more or less reduced or wanting. In the prothorax they are of smaller calibre and are attached to the occipital region of the head.

(*b*) *Ventral longitudinal*; these extend from one sternal apophysis to another. In the prothorax they pass to the head, where they are inserted on the occiput or the tentorium.

(*c*) *Ventral oblique*, consisting of anterior and posterior series that arise from the sternal apophysis and are attached to the spina of the segment in front and behind, respectively.

(*d*) *Dorsoventral* or *tergosternal* muscles which act as the main levators of the wings and are therefore antagonistic to (*a*). They are wanting in the prothorax and reduced in flightless insects. The two series (*a*) and (*d*) are known as the *indirect wing muscles* because their insertions are not upon the wing bases. The next group (*e*) includes what are commonly termed the *direct wing muscles* (see pp. 33–4) on account of their attachments being very near to or on the wing bases.

(*e*) *Pleural muscles* which include (1) the *anterior extensors of the wing* arising from the pleuron and the coxal margin and inserted on the anterior pleural (basalar) wing sclerite, beneath the wing base; (2) the *posterior extensor of the wing*, whose origin is on the margin of the coxa of its segment and the insertion on the posterior pleural (subalar) wing sclerite, beneath the wing base;

(3) the *flexor of the wing*, which arises from the pleural ridge and is inserted on the wing base. A further group of *accessory indirect wing muscles* runs from the tergum or sternum to the pleuron. Their action braces the thoracic exoskeleton or modifies its elastic properties during flight.

(*f*) *Leg muscles*, which comprise extrinsic (Fig. 14) and intrinsic (Fig. 15) series. The *extrinsic series* is concerned with movements of the leg as a whole. Arising from the tergum and inserted on

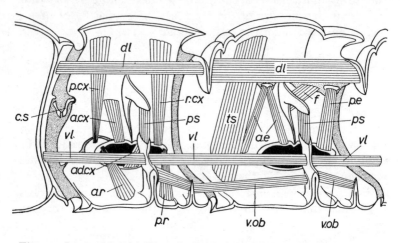

Fig. 14. Internal lateral view of pro- and mesothorax of a typical insect, to illustrate musculature

a.cx, abductor of coxa; *ad.cx*, adductor of coxa; *a.e*, anterior extensors of fore wing; *a.r*, anterior rotator of coxa; *c.s*, cervical sclerites; *dl*, dorsal longitudinal muscles; *f*, flexor of fore wing; *p.cx*, tergal promotor of coxa; *p.e*, posterior extensor of fore wing; *p.r*, posterior rotator of coxa; *ps*, pleurosternal muscle; *r.cx*, tergal remotor of coxa; *ts*, tergosternal muscle; *vl*, ventral longitudinal muscles; *v.ob*, ventral oblique muscles

the trochantin and coxa, respectively, are the *promotor* and *remotor of the coxa*; these muscles effect forward and backward movements in the horizontal plane. An *abductor* and an *adductor of the coxa* bring about up-and-down movements of the leg and arise from the pleuron and sternum respectively. *Anterior* and *posterior rotators* of the coxa, of sternal origin, effect partial rotation of the limb. Finally, the *depressor of the trochanter* (Fig. 15) represents a group of muscles with origins on the tergum, furca and ventral margin of the coxa. When the articulation between the femur and

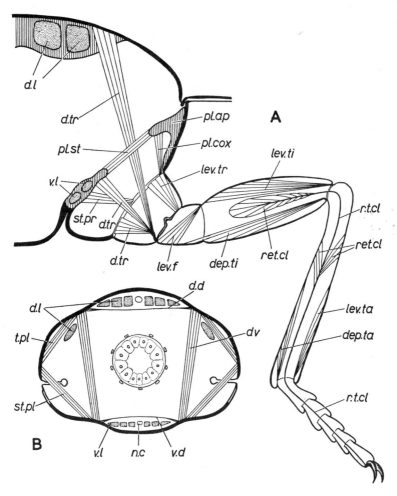

Fig. 15. Skeletal musculature

A. Transverse section through thorax, showing intrinsic and some extrinsic muscles of leg. B. Transverse section through abdomen. *d.d*, dorsal diaphragm; *dep.ta*, depressor of tarsus; *dep.ti*, depressor of tibia; *d.l*, dorsal longitudinal muscles; *d.tr*, depressor of trochanter; *dv*, dorsoventral muscles; *lev.f*, levator of femur; *lev.ta*, levator of tarsus; *lev.ti*, levator of tibia; *lev.tr*, levator of trochanter; *n.c*, nerve cord; *pl.ap*, pleural apophysis; *pl.cox*, pleurocoxal muscle; *pl.st*, pleurosternal muscle; *ret.cl*, retractor of claws; *r.t.cl*, retractor tendon of claws; *st.pl*, sternopleural muscle; *st.pr*, sternal promotor of coxa; *t.pl*, tergopleural muscle; *vd*, ventral diaphragm; *v.l*, ventral longitudinal muscle

trochanter is fixed, this muscle acts as a depressor of the leg.

The *intrinsic series* comprise (1) the *levator of the trochanter* which arises from the base of the coxa and is inserted into the base of the trochanter. As in the case of the depressor, this muscle moves the leg as a whole where the femoro-trochanteral joint is fixed. The only muscle moving the femur is the *levator of the femur* which arises ventrally on the trochanter and is attached to the dorsal tip of the base of the femur: it is wanting in the hind legs of locusts and crickets. The cavity of the femur is largely occupied by the *levator* and *depressor of the tibia*. A single muscle, the *retractor of the claws*, has points of attachment on the femur and tibia and its long tendon is inserted on the unguitractor plate of the pretarsus. Its action is to pull the claws downward and towards the tarsus, the protraction of the claws being effected by the elasticity of their basal supporting parts. Originating in the distal half of the tibia are the *levator* and *depressor of the tarsus*; their insertion is on the dorsal and ventral borders, respectively, of the base of the first tarsomere.

3. **The Abdominal Muscles.** The absence of legs and wings greatly simplifies the myology of the abdomen but there are special muscles associated with the ovipositor and male genitalia. The principal abdominal muscles include the following (Fig. 15 B):

(a) *Dorsal longitudinal*, whose origins and insertions are on the intersegmental folds. They form a series along each side of the heart.

(b) *Ventral longitudinal*, which are the counterparts of (a) and lie on each side of the ventral nerve cord.

(c) *Dorsoventral*, which are mostly tergosternal in their attachments. They either lie within their segments of origin or cross from one segment to the next. They function as compressors of the abdomen and are of importance in respiration (p. 84).

(d) *Pleural*, including tergopleural and sternopleural series; in some insects they function as dilator and occlusor muscles of the spiracles.

THE NERVOUS SYSTEM

The tissue of the nervous system of insects comprises two main types of cells: the nerve cells or *neurons* and the non-nervous, irregularly branched, interstitial cells forming the *neuroglia*. The neurons, which are grouped together in nerve centres or *ganglia*, are greatly attenuated cells derived from ectoderm and specialized for the rapid conduction of the electro-chemical nervous impulses. Each neuron consists of a cell body with its nucleus, and one or more nerve fibres or *axons*. According to the number of axons a neuron is described as *unipolar*, *bipolar* or *multipolar*. The axon often has a side-branch or *collateral*, and both end in delicate branching fibrils – the *terminal arborization*. Each axon is enclosed in a nucleated coat – the *neurilemma* – and though there is no thick myelin sheath of the kind found in vertebrates, the cytoplasm of the axon is surrounded by a thin layer of fat-like material. Three kinds of neurons occur commonly: the *sensory*, *motor* and *associa-tion neurons*, while a fourth type is represented by *neurosecretory cells* with special endocrine functions (p. 97). The sensory neurons are associated with the sense organs and lie near the integument; they are never found within the central nervous system, as in vertebrates. Each sensory neuron is usually bipolar. The distal process is adapted to respond to a particular kind of stimulus while the long axon (which develops as an outgrowth of the neuron) ends ultimately in an arborization within a central ganglion (Fig. 16). Fibres from neighbouring sensory neurons may be grouped to-

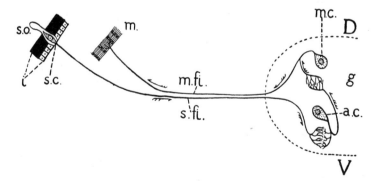

Fig. 16. Diagram of the reflex mechanism of an insect

One-half of a ganglion, *g*, of the ventral nerve cord is shown. D, dorsal. V, ventral. A motor fibre, *m.fi*, and a sensory fibre, *s.fi*, of a lateral nerve are shown. *i*, integument; *s.o*, sense organ; *s.c*, sensory neuron or cell; *m*, muscle; *a.c*, association neuron; *m.c*, motor neuron. (The course traversed by the nerve impulses resulting from stimulation of the sense organ, is shown by arrows)

gether to form a *sensory (afferent) nerve*. The motor neurons always lie within the ganglia; they are mostly unipolar and their axons may be grouped together to form *motor (efferent) nerves*. These pass mostly to the muscles where the axons terminate in minute conical *endplates* or in fine branches on or within the muscle fibres. The association neurons form, with their processes, links between the sensory and motor neurons as shown in Fig. 16. The resulting junctions between the arborizations of adjacent nerve cells are known as *synapses*. The terminal branches of the two neurons are not usually in actual contact at a synapse; each nerve-impulse which arrives there causes the temporary release of a transmitter substance such as acetylcholine or γ-aminobutyric acid which then activates the adjacent neuron, so ensuring transmission of the impulse across the synapse.

The nervous system comprises the central nervous system, the visceral nervous system and the peripheral sensory nervous system.

The Central Nervous System. This is composed of a double series of nervous centres or *ganglia* joined together by longitudinal and transverse tracts of nerve fibres. The longitudinal tracts of fibres are the *connectives* (see Fig. 17) and they serve to unite a pair of ganglia with those in front and behind. The transverse fibres or *commissures* unite the two ganglia of a pair. While there is usually a pair of ganglia in almost every segment in the lower insects (Fig. 17 A), a varying degree of fusion occurs in the higher groups (C, D). Also, the members of a pair are often so closely amalgamated that they seem to form a single ganglion. The connectives are separate and distinct in the more primitive insects, but often they are so closely approximated as to appear as a single longitudinal cord (Fig. 17 D).

The ganglia of the central nervous system are mainly formed of peripheral aggregations of nerve cells enclosing a central mass of nerve fibres (the so-called *neuropile*). Each lateral nerve has two roots. The fibres of the dorsal root arise from motor cells situated dorsolaterally in a ganglion and the sensory fibres, composing the ventral root, end in terminal arborizations on the ventrolateral aspect of the ganglion. The association neurons lie for the most part between the dorsal and ventral roots. Externally the ganglia and nerves are invested with a sheath or *neurilemma*, secreted by

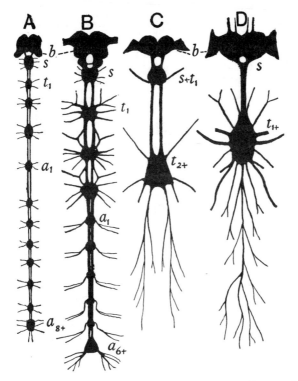

Fig. 17. Types of nervous system (adapted from various authors)
A *Japyx*. B, *Blatta*. C, Water bug (*Belostoma*). D, Housefly (*Musca*). *a* abdominal ganglia; *b*, brain; *s*, suboesophageal ganglion; *t*, thoracic ganglia. The subscript numbers denote the segment supplied by the ganglia

a superficial layer of cells, the *perineurium*, which controls differences in the ionic composition of the blood and the nervous tissue. The central nervous system is divisible into the brain and the ventral nerve cord.

The Brain (or *supra-oesophageal ganglia*). The brain (Figs. 17 and 18) lies just above the oesophagus and is formed by the amalgamation of the ganglia of the presegmental acron and three anterior head segments. They develop very unequally and give rise to the *protocerebrum*, the *deutocerebrum* and the *tritocerebrum*. The *protocerebrum* forms the greater part of the brain and represents the ganglia of the preantennal segment and the acron; it innervates

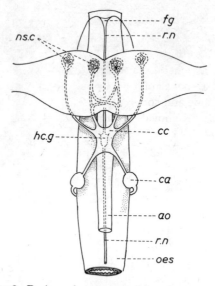

Fig. 18. Brain and stomatogastric nervous system

ao, aorta; *ca*, corpus allatum; *cc*, corpus cardiacum; *fg*, frontal ganglion; *hc.g*, hypocerebral ganglion; *ns.c*, neurosecretory cells of brain; *oes*, oesophagus; *r.n*, recurrent nerve

the compound eyes and ocelli. Laterally this region is expanded to form the *optic lobes* which contain the elaborate system of nerve cells and fibres linking the visual cells of the eye with the centres of the protocerebrum. Their degree of development is correlated with that of the eyes. Within the protocerebrum are the so-called *mushroom bodies*, composed of vast numbers of small association cells whose axons are grouped together into stalk-like bundles. The size and complexity of these bodies corresponds in a general way with complexity and specialization of behaviour. The *deutocerebrum* is formed by the ganglia of the antennal segment: it is chiefly composed of the antennal lobes and innervates the antennae and their muscles. The *tritocerebrum* is formed by the third pair of segmental ganglia and lies beneath the antennal lobes. It is poorly developed owing to the absence of the second antennae in insects and its function is to innervate the labrum and the fore intestine. Its component ganglia are far apart and are joined by the *postoesophageal commissure*.

The brain contains a few motor neurons concerned with

antennal movements but its main functions are those of sensation and coordination. It is responsible for maintaining the general tonus of the skeletal muscles, it controls the local reflexes which are mediated by the thoracic and abdominal ganglia, and it exerts an inhibitory action on centres in the suboesophageal ganglion. Experimental removal of one side of the brain reduces muscle tonus on that side and so results in circus movements towards the uninjured side. Removal of the whole brain allows the suboesophageal ganglion to excite the locomotor centres and so causes the insect to walk restlessly in response to slight stimuli.

The Ventral Nerve cord. The ventral nerve cord (Fig. 17) is a median chain of segmental ganglia lying beneath the alimentary canal. It is joined to the tritocerebrum by the *para-oesophageal connectives*. The first ventral nerve centre is the *suboesophageal ganglion* formed by the fusion of the fourth to sixth neuromeres or ganglia of the mandibular, maxillary and labial segments. It gives off paired nerves supplying their respective appendages. There follow three thoracic ganglia and, at most, eight ganglia in the abdomen. The first abdominal ganglion often fuses with that of the metathorax, and the end ganglion of the chain is always a composite centre formed by the fusion of at least three neuromeres. In some groups of insects extensive fusion of the ventral ganglia occurs, especially in the Hemiptera and higher Diptera (Fig. 17 C and D). In extreme cases all the ventral ganglia (including the suboesophageal) are amalgamated into a large compound ganglion from which nerves run to all parts of the trunk. The thoracic ganglia innervate the legs and wings, while each abdominal ganglion shows considerable autonomy and functions to some extent as a local centre for its segment. A complicated reflex act such as oviposition can be carried out by a living isolated abdomen when suitably stimulated, provided the last ganglion and its nerves are intact.

The Visceral Nervous System. The principal component of the visceral or sympathetic nervous system is that known as the *stomatogastric system* (Fig. 18), which is formed by ingrowth of the dorsal part of the stomodaeum. It includes, firstly, a median *frontal ganglion* (*fg*) lying just anterior to the brain. This ganglion seems to exert an effect on the ionic concentrations of the blood; it is joined by bilateral connectives to the tritocerebrum and gives off a *recurrent nerve* (*rn*) that ends in the *hypocerebral ganglion* (*hg*). Behind the brain there are also paired *corpora cardiaca* (*cc*) that are joined to the protocerebrum and the hypocerebral ganglion. The corpora cardiaca exert important endocrine functions and closely associated with them are further paired endocrine glands, the *corpora allata* (*ca*: p. 98). A single or paired *ventricular ganglion* is linked to the hypocerebral ganglion. The stomatogastric system contains both motor and sensory fibres and innervates the heart and fore intestine.

The *ventral sympathetic system* consists of transverse nerves associated with each ganglion of the ventral cord; they supply the spiracles of their segment. Arising from the last abdominal ganglion are *splanchnic nerves* that innervate the reproductive organs and the hind intestine.

The Peripheral Sensory Nervous System. This is composed of a fine network of axons and sensory cells lying beneath the integument. The nerve cells have branched distal processes that end in the epidermis itself. The axons combine and enter the paired segmental nerves of the ventral cord. This system is perhaps homologous with the nerve net of the lower invertebrates. Among insects it is best developed in soft-skinned larvae.

THE SENSE ORGANS AND PERCEPTION

Sensory perception is achieved by means of structures termed *receptors* or *sensilla*. These take various forms and are situated at the peripheral endings of the sensory nerves. In many cases – the tactile receptors for example – they are scattered in distribution, whereas in the eyes and tympanal organs they are aggregated, often in large numbers. In their least modified form receptors closely resemble ordinary body hairs and only differ in being connected with the nervous system. The components of a simple type of receptor (Fig. 19 A) are the cuticular or external part with its trichogen or formative cell together with a bipolar sense cell. The latter lies in or just beneath the epidermis and its distal process penetrates the trichogen cell to enter the cavity of the cuticular part of the receptor. In many cases a tormogen or membrane cell is also present. Various types of receptors are evidently derived from this simple trichoid structure. They are characterized by the form of the cuticular parts and may be *basiconic, placoid, campaniform, coeloconic,* etc., that is, peg-like, plate-like, dome-like or in pits. Receptors of a different kind are the ommatidia or components of the eyes, and the chordotonal receptors of the tympanal organs.

Not all the receptors have yet had functions ascribed to them with certainty, though their probable role may sometimes be inferred from their structure and position. Further information has been gained by studying the reactions of insects in which receptors, or parts of the body bearing them, have been removed or coated, for example, with impermeable substances. More recently, electrical methods of recording impulses in the sensory nerves, or even within the receptor cells, have provided more exact physiological knowledge of their functions. In many cases, however, the minute size of the sensilla and the fact that several kinds often occur in close proximity has made their investigation difficult.

The following classification of receptors is convenient:

(*a*) Mechanoreceptors, e.g., of touch, tension and balance.
(*b*) Auditory organs, perceiving sound.
(*c*) Chemoreceptors, perceiving odours and tastes.
(*d*) Temperature and humidity receptors.
(*e*) Photoreceptors or visual organs.

The majority of these are *exteroceptors*, perceiving stimuli which arise outside the insect. Some mechanoreceptors, however, serve as *proprioceptors*, i.e., they respond to internal stimuli resulting from changes in the position of the body or its parts.

Mechanoreceptors. These are excited by stimuli which temporarily deform the cuticle in or near the sensillum; three main types occur. In the articulated *sensory hairs* (Fig. 19 A) movement of the hair in its socket produces impulses in the sensory nerve fibre. Such sensilla mediate the sense of touch or perceive currents of air or water. When grouped together in the form of 'hair plates' near the joints of an appendage they act as proprioceptors, since they are then stimulated when one segment of the limb moves against its neighbour. The *campaniform sensilla* (Fig. 19 B) consist of minute circular or oval dome-shaped areas of thin cuticle, each of which is in contact internally with the rod-like terminal process

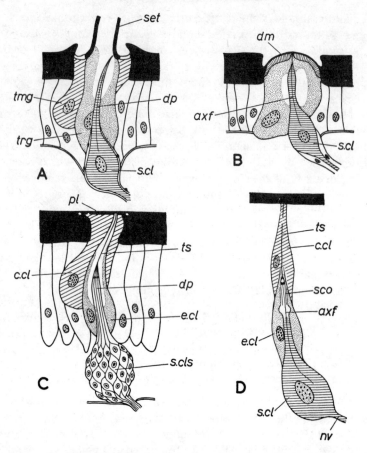

Fig. 19. Sensory receptors (*after* Snodgrass)

A. Sensory hair. B. Campaniform sensillum. C. Placoid sensillum. D. Chordotonal sensillum. *axf*, axial fibre; *c.cl*, cap cell; *dm*, cuticular dome; *dp*, distal process of sense cell; *e.cl*, envelope cell; *nv*, axon of sense cell; *pl*, cuticular plate; *s.cl(s)*,. sense cell(s); *sco*, scolopale; *set*, base of seta; *tmg*, tormogen cell; *trg*, trichogen cell; *ts*, terminal strand

or *scolopale* of a sensory nerve cell. When movement of the body causes stresses to develop in the adjacent cuticle, the scolopale is displaced up or down and so stimulates the neuron to discharge impulses in the sensory fibre. Campaniform sensilla are thus essentially proprioceptors and are commonly found in groups near the joints of the legs and palps and at the bases of the wings and halteres. The *chordotonal sensilla* (Fig. 19 D) occur singly or grouped in many parts of the body of insects. In the grasshopper *Melanoplus*, for example, Slifer found 76 pairs of chordotonal organs, each composed of one or more of these receptors. A chordotonal sensillum consists of a long cap cell attached to the integument, an envelope cell and a sense cell. The envelope cell surrounds the distal process of the sense cell, whose apex is prolonged into a terminal fibre that is fastened to the cuticle. At the apex of the sensory process is a sense rod or scolopale of complex structure. A delicate axial fibre arises in the sense cell and traverses the scolopale to end in its deeply staining cap or apical body. Any displacement of the scolopale, it would appear, excites the sense cell through the axial fibre. Very often chordotonal receptors are attached at both their ends to the integument, and it is probable that they are proprioceptors sensitive to changes of tension in the muscles. Intramuscular proprioceptors not unlike those of vertebrates have been found in some insects, but appear to be less important than the chordotonal and campaniform sensilla.

Specialized organs of equilibrium occur in some insects; they depend on the presence of one or more of the above types of mechanoreceptors. The halteres of Diptera, for example (pp. 36 and 185), which are concerned with the maintenance of equilibrium in flight, are provided basally with specially arranged groups of campaniform sensilla. These register the alterations in cuticular stresses at the base of the vibrating haltere when the insect changes its direction of flight. In the second antennal segment (pedicel) of almost all insects lies *Johnston's organ*. It is formed of chordotonal sensilla attached distally to the articular membrane at the base of the flagellum and thus perceives the latter's movements. In the Chironamidae and Culicidae, where the organ is highly developed and has very many sensilla, the pedicel is specially enlarged to accommodate it.

Auditory Receptors. Sound waves transmitted through the air are perceived by specialized organs which respond to very small displacements of their cuticular parts and are thus related to the mechano receptors described above. In male Culicidae, Johnston's organ acts as a sound receptor since the densely plumose flagellum is moved by the sound waves, while auditory hairs, not very different in structure from the tactile ones, occur on the cerci of cockroaches and crickets. But the most specialized sound receptors are the *tympanal organs* (Fig. 20 C, E) such as occur on each side of the first abdominal segment in the Acrididae (short-horned grasshoppers), or on the proximal part of the fore tibiae in Tettigoniidae (long-horned grasshoppers) and Gryllidae (crickets). Comparable organs are found also in cicadas and many moths. In all cases they have an essentially similar structure. A delicate external cuticular membrane, the *tympanum*, overlies a tracheal

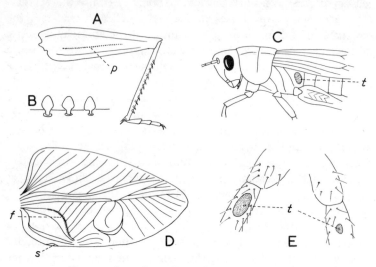

Fig. 20. Sound-producing and auditory organs

A, hind leg of an Acridid grasshopper showing row of stridulatory pegs (*p*); B, three pegs at higher magnification; C, lateral view of an Acridid grasshopper showing tympanal organ (*t*) on first abdominal segment; D, fore wing of male cricket (*Acheta*), showing stridulatory file (*f*) and scraper (*s*); E, outer, and F, inner, views of tympanal organ (*t*) on fore tibia of cricket (*Acheta*)

air-sac and is therefore able to vibrate freely when sound waves impinge on it. The movements of the tympanum then stimulate the chordotonal sensilla attached to it. The tympanal organs of Orthoptera are most sensitive to sounds with frequencies between about 5000 and 20 000 cycles per second while those of moths are sensitive to ultrasonic frequencies. In some cases it has been demonstrated that the responses in the tympanal nerves are related to the frequency with which the amplitude of the sound is modulated; the fundamental frequency of the 'carrier wave' is important only in that it must fall within the auditory range of the insect. The latter therefore distinguishes sounds which differ in the rhythmic variations of intensity which make up the pattern of amplitude modulation; in this respect it differs greatly from man, who relies mainly on differences in pitch (frequency), though some degree of pitch discrimination is also found in insects.

Correlated with the ability of many insects to perceive sounds is their ability to produce noises and therefore to communicate with each other over relatively great distances. The sounds are produced in various ways. Thus, the familiar stridulation of many short-horned grasshoppers is caused by rubbing the inner surface of each hind femur – which bears a row of minute pegs – against a thickened vein of the adjacent closed tegmen (Fig. 20 A, B). In the crickets (Fig. 20 D), each tegmen bears a rasping organ or *file* and a hardened area or *scraper*, the file of one tegmen working against the scraper of the other and so throwing into vibration the specialized resonating areas of the tegmen. In the long-horned grasshoppers there is a similar mechanism but the file is only functional on the left tegmen and the scraper on the right one. Male cicadas have a pair of drum-like *tymbals* at the base of the abdomen and loud sounds are produced when the 'head' of the drum is repeatedly moved in and out through the frequent action of powerful muscles, rather in the way that sounds can be made by pushing the lid of a tin can in and out. In all the above cases sound production is confined to or predominates in the males, the sounds may have relatively elaborate patterns, they display amplitude modulation and their characteristics are recognized by other members of the species through the auditory organs. A given species may produce more than one kind of song and at least some of these play an important role in attracting or stimulating the female in courtship. Many other insects produce sounds by a great variety of means, but are not equipped with specialized auditory organs; in these cases the sounds may be defensive or of unknown significance.

Chemoreceptors. In this category are the receptors for smell and taste. In terrestrial insects the olfactory sense is stimulated by low concentrations of the vapours of volatile substances, while the gustatory sense perceives relatively high concentrations of the stimulant in aqueous solution. Whether a comparable distinction holds good for insects living in wet environments is not clear.

The olfactory receptors are found mainly on the antennae and less often on the palps. Each receptor possesses cuticle which is at least partly very thin and perforated by minute pores, and which is usually innervated by a group of bipolar neurons whose distal processes have a ciliary ultrastructure. The various different structural types of receptors have been classified as *trichoid, basiconic, coeloconic* and *placoid* (Figs. 19 C, 21). They are usually scattered over the surface of the organ bearing them, but are sometimes grouped in pits to form a definite olfactory organ (e.g., on the antennae of Muscid flies). The receptors may be more numerous in the male, and in *Apis*, for example, the last eight

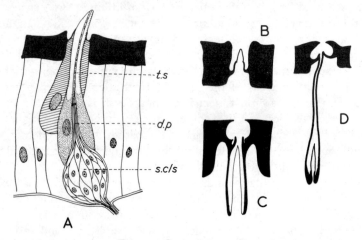

Fig. 21. Cuticular sensilla.

A, basiconic sensillum; B, coeloconic sensillum; C and D, two types of sensilla ampullacea. *d.p*, distal processes of sense cells; *s.cls*, sense cells; *t.s*, terminal strand. In B, C and D only the cuticular structures are shown

antennal segments of males bear about 30 000 placoid sensilla, as compared with 6000 in the workers and 2000 in the queens. The sense of smell plays an important part in the life of insects since many behavioural and developmental changes are caused by *pheromones*. These are highly specific, volatile substances usually perceived through the olfactory sensilla after having been secreted by other members of the same species. Thus, the scents produced by the virgin females of some species of moths attract the males, which may assemble near the female from distances of 3 km or more. Bombykol, the sex-attractant substance of the silk-moth

Bombyx mori, and several other pheromones have now been identified chemically and shown to induce responses at remarkably low concentrations. Social insects, such as bees, wasps and ants, recognize members of their species or colony by smell, and some ants follow odour trails left by their fellows in journeys to and from the nest. Alarm pheromones occur in many species of ants, causing them to become more active or aggressive, while morphogenetic pheromones probably control the differentiation of the various castes in the termite *Kalotermes*. Many female insects are attracted by smell to sites suitable for egg-laying and the development of the newly hatched young while some phytophagous insects are attracted to their host-plants by the smell of the essential oils there.

The *gustatory receptors* are probably basiconic or trichoid. They occur on the surfaces of the pre-oral food cavity and mouthparts, on the antennae of some Hymenoptera and the tarsi of many Lepidoptera, Diptera and honey bees. The Red Admiral butterfly (*Vanessa atalanta*) can, with its tarsal receptors, distinguish between distilled water and a $M/12\,800$ solution of sucrose, a sensitivity over 200 times that of the human tongue. In general, there are wide differences in the taste thresholds of different substances with a given species and for different species with the same substance. Solutions of sugars and dilute solutions of acids and salts are usually preferred to distilled water while more concentrated solutions of acids, salts, esters, alcohols and amino acids are usually rejected. Detailed investigations have shown that in some insects the individual sense cells of a single gustatory sensillum respond specifically to particular substances, some reacting to sugars, others to salts or to water.

Temperature and Humidity Receptors. Temperature receptors may occur on the antennae, maxillary palps and tarsi, though relatively little is known of their structure and physiology. A few insects are known to respond to radiant heat, but more often it seems that convective transfer of heat in the air is the effective stimulus. Blood-sucking and ectoparasitic insects may in this way detect the presence of their warm-blooded hosts while insects given the opportunity of moving freely in an experimental temperature gradient tend to congregate in a zone representing their temperature preference.

Many insects react to differences in atmospheric humidity,

either by showing a preference for certain humidities or by orienting themselves to the vapour from a distant source of water. Basiconic, trichoid and placoid sensilla mediate these responses but their mode of action is not known.

Photoreceptors. Photoreceptors in insects are of two kinds: ocelli or simple eyes, and compound or faceted eyes. Typically both kinds of eyes are found in the same insects, but either or both may be absent.

(a) Ocelli

The most obvious distinction between an ocellus and a compound eye is the presence of a single corneal lens in the former, whereas in the compound eye there are many. Ocelli fall into two classes: the *dorsal ocelli* of imagines and nymphs, and the *lateral ocelli* of most Endopterygote larvae.

Dorsal ocelli are typically three in number arranged in a triangle either on the frontal region of the head or on the vertex (Fig. 3). That the median ocellus of the group was originally paired is shown by its bilateral form in some insects and by the double nerve roots. The dorsal ocelli are innervated from the ocellar lobes of the brain which are located in the protocerebrum between the mushroom bodies. Structurally an ocellus consists of a biconvex *lens* beneath which is a transparent *corneagen layer* which overlies the sensory elements or *retinulae* (Fig. 22). A retinula is composed of a group of visual cells whose contiguous regions are specialized to form a more or less rod-like *rhabdom*, with each cell produced into a fibre of the ocellar nerve. Between the retinulae and around the margin of the lens there are usually *pigment cells*.

Little is known of the function of dorsal ocelli. Their structure shows that they are incapable of any but the crudest kind of image formation, but they probably perceive changes in light intensity. In some insects the response to light received through the compound eyes is more acute and lasting when the dorsal ocelli are in their normal condition than when painted over with an opaque substance. Evidence of this kind suggests that the ocelli and compound eyes interact in mediating the insect's response to light, but the ocelli seem to be more than simple 'stimulatory organs' and

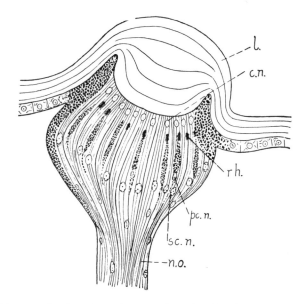

Fig. 22. Section through an ocellus of *Aphrophora spumaria*
c.n, nucleus of corneagen cell; *l*, lens; *n.o*, ocellar nerve; *pc.n*, nucleus of pigment cell; *rh*, rhabdom; *sc.n*, nucleus of retinulae. (*After* Link)

the interaction between eyes and ocelli may sometimes be antagonistic.

Lateral ocelli have no general uniformity of structure and may occur singly (sawfly larvae) or in groups (larvae of Lepidoptera and other orders) on either side of the head. They are innervated from the optic lobes of the brain and when the larva undergoes metamorphosis they are replaced by compound eyes. In Lepidopterous larvae they have the structure of an ommatidium of a compound eye; a corneal lens, often with a crystalline cone, overlies a group of visual cells forming a retinula that secretes a central rhabdom. In the larvae of sawflies, and of some beetles, their structure is similar to that of a dorsal ocellus and comprises a number of retinulae in association with a single lens. Lateral ocelli are able to perceive movements of objects in their vicinity, besides being responsible for orientation to light and perhaps for some appreciation of colour.

(b) Compound eyes (Fig. 23)

Compound eyes are the principal visual organs and are characterized externally by the cornea, or investing cuticle, being divided into facets or lenses, usually hexagonal in form. Each facet is part of a separate visual structure or *ommatidium*. The number of facets (and therefore of ommatidia) to a compound eye ranges from over 20 000 in some dragonflies to 12 000 or more among Lepidoptera and 4000 in *Musca*, while in the workers of certain ants they may be reduced to fewer than a dozen. The cornea is formed of transparent cuticle and is shed at each moult. It is produced by two epidermal cells which form the *corneagen layer*. Between the latter and the visual cells is a group of four transparent *cone cells* which either contain fluid or secrete a body known as the *crystalline cone*. The retinula is composed of a group of usually 7 pigmented *visual cells* which are collectively modified where they are in contact to form an internal *rhabdom*. The ommatidia are isolated from their fellows by pigment cells arranged in two groups: one group surrounds the cone and forms the *primary iris cells* while the other surrounds the retinula and forms the *secondary iris cells*.

In general, the compound eyes can perceive something of the form, movement and spatial location of external objects and detect some differences in the intensity and colour of light falling on them. The formation of an image is accounted for by the mosaic theory of vision. Light rays from external objects are focused on to the rhabdoms by the cornea and cones in such a way that each rhabdom is stimulated by a very small zone of light from that part of the visual field which it subtends. The small zones will normally differ in intensity and together they give rise to an erect image composed of light and dark spots, rather like a newspaper photograph. The process by which light rays are focused on to the rhabdoms is very complicated; it differs from one insect species to another and depends on whether the eye is in a light- or dark-adapted state. The many possible mechanisms have not yet been sufficiently investigated to allow simple modern generalizations, but the classical work of Exner distinguished two principal methods of image formation (Fig. 24). In eyes which form an *apposition image* (Fig. 24 A), each ommatidium is optically separated from its neighbours by dense pigment. Each rhabdom is therefore stimulated only by light rays focused through the dioptric unit

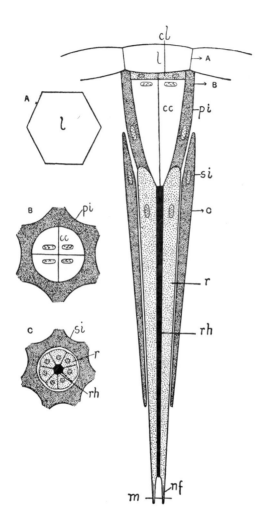

Fig. 23. Diagram of generalized ommatidium of an eucone eye

cc, crystalline cone; *cl*, corneagen layer; *l*, corneal lens; *m*, fenestrated membrane; *nf*, nerve fibre; *pi*, primary iris cell; *r*, retinula; *rh*, rhabdom; *si*, secondary iris cells. A, B and C, transverse sections of regions bearing corresponding lettering

(cornea and cone) immediately above it, and any rays which cannot be focused on to this rhabdom are prevented by the pigment from stimulating adjacent ones. In the so-called 'clear zone eyes', on the other hand, there is a more or less wide, pigment-free gap between the inner ends of the cones and the outer ends of the rhabdoms, at least in the dark-adapted condition. In such eyes, image formation seems to occur by several different methods, one of which involves the formation of a *superposition image*. Here pigment does not separate adjacent ommatidia, with the result that

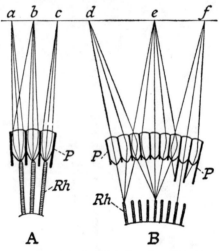

Fig. 24. Diagram showing classical theories of image formation by the compound eye

A, eye forming apposition image. B, eye forming superposition image. *a–f*, luminous points with the course of the rays emitted by them; *P*, pigment; *Rh*, rhabdom. At the right side the migration of pigment into the light-adapted position results in an apposition image; all rays except those entering the central facet are intercepted. (*From* Wigglesworth *after* Kühn)

light rays can be brought to a focus on a given rhabdom after passing through the dioptric units of several adjacent ommatidia, as shown in Fig. 24 B. The superposition image is less distinct, but brighter, since less light is wasted through absorption by pigment. This type of eye is therefore often found in nocturnal or crepuscular insects, while diurnal insects usually form an image by apposition. In some species, however, migration of the pigment enables the eye to function in either way, depending on light intensity, so that the phenomenon of adaptation occurs (see the right-hand side of Fig. 24 B). The clear zone may also be crossed by crystalline tracts which guide light within an ommatidium from the cone to the rhabdom.

Though insects readily detect the movement of objects, the acuity of their vision is much less than that of man. It depends, among other things, on the number and angular separation of ommatidia in the eye but it is too low to make accurate recognition of form possible. Perception of form is, in fact, an elaborate process and the honey bee's ability to distinguish between static patterns of different complexity seems to depend on the changes of stimulation which occur as the image passes over the retina of the moving insect. Some, but not all, insects have colour vision, though different species may distinguish between different sets of colours and many insects are known to perceive ultra-violet radiation, which is invisible to man. By training honey bees to associate certain colours with the presence of food it has been found that they distinguish between six 'colours'. One of these is in the ultra-violet, others correspond to our violet, blue, blue-green and yellow while the last is 'bee purple', a mixture of yellow and ultra-violet. They do not discriminate between the different hues within these 'colours'. A number of insects are able to perceive the plane of vibration of polarized light; honey bees foraging for nectar under a partially clouded sky use this faculty for direction finding.

ALIMENTARY CANAL, NUTRITION AND DIGESTION

(a) The Alimentary Canal

The alimentary canal (Fig. 25) is divided into three regions, the *fore intestine* that arises in the embryo as an anterior ectodermal ingrowth (stomodaeum); the *hind intestine* that arises as a similar posterior ingrowth (proctodaeum); and the *mid intestine, stomach* or *ventriculus* formed as an endodermal sac (mesenteron) connecting the two. These differences in embryonic origin result in marked histological differences in the structure of the mid intestine as compared with either of the other regions. The fore and hind intestine, being ingrowths of the integument, resemble the latter histologically and are lined with cuticle.

Pre-oral Food Cavity. Strictly speaking, this is not part of the alimentary canal, but may conveniently be discussed here. In insects with simple biting mouthparts and a hypognathous head it is the space bounded in front by the inner surface of the labrum, behind by the labium and laterally by the mandibles and maxillae. The hypopharynx (p. 24), which arises near the base of the labium, lies within this cavity and partially divides it into an anterior *cibarium* and a posterior *salivarium*. The cibarium is provided with dilator muscles which arise on the postclypeus and in some insects with specialized feeding habits it forms the *cibarial sucking pump* (e.g., Hemiptera). The salivarium receives the common duct of the labial glands. These normally secrete saliva and in the Hemiptera the salivarium is modified into a *salivary syringe*

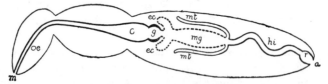

Fig. 25. Diagram of the digestive system of an insect

The ectodermal parts are represented by heavy lines and the endodermal parts by broken lines. *m*, mouth; *oe*, oesophagus; *c*, crop; *g*, gizzard; *ec*, enteric caeca; *mg*, mid intestine; *mt*, Malpighian tubules (often regarded as endodermal); *hi*, hind intestine; *r*, rectum; *a*, anus

which helps to inject saliva into the plant or animal on which the insect is feeding. In Lepidopterous larvae, where the labial glands secrete silk, the salivarium forms the *silk regulator*.

Fore Intestine. This begins with the *pharynx*, a relatively narrow tube which leads back from the pre-oral food cavity and is equipped with dilator muscles arising from the frons. In some insects, such as the Lepidoptera and Hymenoptera, the pharynx may participate in the formation of a sucking pump. Posteriorly the pharynx passes into the *oesophagus*. This may be a simple tube leading into the mid intestine or it may be expanded at some point to form a *crop*. The latter may be a simple dilation of the posterior part of the oesophagus (Fig. 27) or, as in the blow-fly and most Lepidoptera, it is a lateral diverticulum connected with the oesophagus by a narrow tube, illustrated in Fig. 28. The crop acts mainly as a food reservoir, from which the food is transferred to the mid intestine as required, but some digestion may also take place there. The fore intestine is separated from the mid intestine by the cardiac sphincter, and at this region it is often modified to form a muscular *proventriculus* or *gizzard* which varies greatly in its development. Its main function is that of a sieve occluding the passage of food if not in a sufficiently divided state. Well-developed in many chewing insects, in sucking insects it is little more than a valve. In the cockroach and some Orthoptera and Coleoptera it has powerful radial teeth and circular muscles and serves the additional function of crushing the larger food particles.

Histologically, the fore intestine consists of a cellular layer which secretes the cuticular lining and is covered externally with a basement membrane. Outside the latter is a coat of longitudinal muscle fibres overlaid by circular fibres.

The **Mid Intestine** is composed of a layer of large epithelial cells bounded externally by a basement membrane. On their inner aspect these cells usually show a *striated border* (Fig. 26) composed of large numbers of *microvilli*. A layer of minute circular muscle fibres and external longitudinal muscles is present. The superficial area of the mid intestine is increased not only by the microvilli but also, in many insects, through the development of outgrowths or *caeca* that vary in size and number (Fig. 27). A third method of achieving the same result is by the folding of the epithelial layer to form crypts: all methods may occur in the same species. The epithelial cells of most insects usually appear to be of one kind, although individual cells may be in different phases at a given time. Secretion and absorption are therefore thought to take place by the same or similar cells. The method of secretion may be (*a*) *merocrine*, in which the cells discharge their products through the striated border without undergoing any drastic changes; and (*b*) *holocrine*, which is less prevalent but occurs in Orthoptera and is characterized by the disintegration of the cells during the process. Larval Lepidoptera and Trichoptera are unusual in having a second type of midgut epithelial cell, the so-called *goblet cell*, with a large central vacuole. Between the bases of the epithelial cells are small groups of *replacement cells* which divide and provide new epithelial cells (Fig. 26).

In most insects and their larvae the food is separated from the epithelial lining by the *peritrophic membrane* (Fig. 26) which forms a thin, colourless tube projecting backwards into the hind intestine. It is composed partly of chitin and is supposed to protect the epithelial cells from abrasion. This explanation agrees with the absence of mucous cells that might otherwise perform this function. Also, the membrane is absent or greatly reduced and diffuse in many, but not all, insects that imbibe fluid food such as Hemiptera, adult Lepidoptera and numerous blood-sucking forms. The peritrophic membrane is shed through the anus with the faeces. It is usually formed as a secretion from the surface of the mid-intestinal epithelium, but in some insects it is extruded in tubular form by an annular press formed from part of the cardiac valve. It is permeable to digestive enzymes and the products of digestion.

In the larvae of most Plannipennian Neuroptera and Apocritan Hymenoptera the capacious mid gut is closed posteriorly until the end of larval life; communication with the hind gut is then established and the accumulated stomach contents pass out.

The **Hind Intestine** is sometimes divisible into a narrow anterior

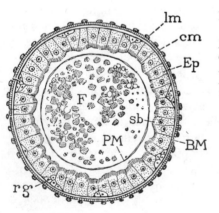

Fig. 26. Diagrammatic cross-section of mid intestine

BM, basement membrane; *cm*, circular muscles; *Ep*, epithelium; *F*, food; *lm*, longitudinal muscles; *PM*, peritrophic membrane; *rg*, regenerative cells; *sb*, striated border. (*From* Snodgrass)

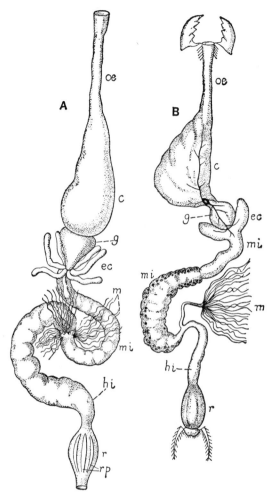

Fig. 27. A, Alimentary canal of *Periplaneta americana*. B, Alimentary canal of *Nemobius sylvestris* (Gryllidae)

oe, oesophagus; *c*, crop; *g*, gizzard; *ec*, enteric caeca; *m*, Malpighian tubes; *mi*, mid intestine; *hi*, hind intestine; *r*, rectum; *rp*, rectal papillae. (*After* Bordas)

tube or *ileum*, a *colon*, and a wider end region or *rectum* which opens exteriorly at the *anus*. Histologically, the hind intestine consists of the same layers as the fore intestine, but the cuticular lining is thinner and the circular muscles are present both external and internal to the longitudinal layer. In many insects the cellular layer

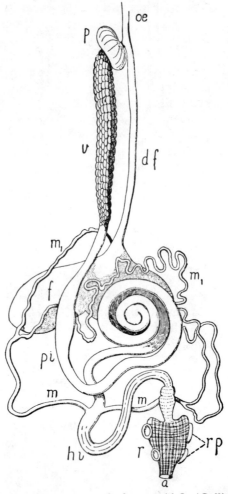

Fig. 28. Alimentary canal of a muscid fly (*Calliphora*)

oe, oesophagus; *p*, proventriculus; *v*, ventriculus; *df*, duct of food reservoir, *f*; *pi*, proximal intestine; *m₁*, Malpighian tubes which unite to form a common stem (*m*) on either side; *hi*, hind intestine; *r*, rectum; *rp*, rectal papillae; *a*, anus. (*Adapted from* Lowne)

of the rectum becomes greatly thickened to form six longitudinal pads or *rectal papillae*. An important function of these papillae is the maintenance of the proper salt and water balance by absorption from the contents of the rectum (see p. 93). A *pyloric sphincter* separates the stomach from the hind intestine and, close to the junction of these two regions, there are outgrowths known as *Malpighian tubules* (p. 93). These vessels are the principal organs of excretion and are slender blind tubes lying in the haemocoele where they are freely bathed in blood. They vary greatly in number, ranging, for example, from 30 to 120 in Orthoptera and from 4 to 6 among many Endopterygota; in the Aphididae and the Collembola these organs are wanting. Each tubule is composed of large epithelial cells resting on an external basement membrane; outside the latter there are commonly muscle fibres. The inner margins of the cells show a *striated border* resembling that of the mid gut and there is no cuticular lining.

Salivary Glands are paired glands which discharge their secretion into the pre-oral food cavity, where it mixes with the food as this is being taken in. Usually these organs are labial glands which originate as paired invaginations of the ectoderm in close association with the developing labial rudiments. Their main ducts fuse to form a common outlet discharging on or near the hypopharynx. In many insects the cuticular lining of the ducts is spirally thickened as in tracheae. In caterpillars the labial glands are converted into silk-producing organs and their salivary functions are assumed by the mandibular glands.

(b) Digestion

Before it can be absorbed by the mid gut, the insect's food usually has to be broken down into simpler soluble substances, such as the monosaccharide sugars and amino acids. These changes are accomplished with the assistance of digestive enzymes secreted by the salivary glands and mid-gut epithelium, digestion occurring in the crop and mid gut. Three main groups of enzymes have been found in insects: (*a*) the *carbohydrases*, which catalyse the breakdown of complex carbohydrates to simple sugars. The amylases of the saliva and mid gut act on starch, while the glycosidases of the mid gut control the breakdown of the complex sugars like maltose, sucrose and lactose; (*b*) the *lipases*, which catalyse the breakdown of fats; (*c*) the *proteases*, which are responsible for the digestion of proteins. The endopeptidases act on proteins or peptones, converting them to polypeptides, while the exopeptidases complete digestion by breaking down the peptides into amino acids.

In *Glossina* and some other blood-sucking insects the proteases are abundant, but the carbohydrases are almost absent. On the other hand, in butterflies and moths – which feed mainly on nectar – almost the only enzymes present are the invertases which hydrolyse cane-sugar. No plant-feeding insects can digest lignin and most of them cannot even make use of the cellulose in their diet. Of those insects which can utilize cellulose, a few species secrete cellulose-splitting enzymes (e.g., Cerambycid beetle larvae); more commonly, however, the breakdown of cellulose is accomplished by symbionts living in the gut. These may be bacteria, as in some Scarabaeid larvae, or Protozoa, as in the wood-feeding cockroach *Cryptocercus* and the more primitive termites. While the symbionts are present, such termites can live for long periods on pure cellulose, but they soon die if deprived of the Protozoa experimentally (e.g., by exposing the insects to $36°$ C for 24 hours or to an oxygen tension of 3–4 atmospheres). The Protozoa are not transmitted directly to the progeny and are lost at each moult with the other contents of the gut. Newly-hatched or newly-moulted termites acquire a new protozoal fauna by feeding on material exuding from the anus of other members of the colony.

(c) Nutrition

The basic nutritional requirements of a number of species are now known in biochemical terms. Carbohydrates are the main source of energy, though proteins and fats may also be oxidized for this purpose. The amino acid requirements of a few species are known in detail and it has been shown that some amino acids are essential for growth and development. Most insects require an external source of valine, arginine, histidine, tryptophan, leucine, *iso*-leucine, lysine, methionine, phenylalanine and threonine. The diet must also include certain vitamins; different species have different needs but a sterol and many of the B-complex vitamins are usually essential. In some cases these vitamins are not consumed in the food but are manufactured within the insect by symbiotic micro-organisms. These may live in the gut, e.g., the yeast-like *Nocardia rhodnii* of *Rhodnius*, or they are lodged in special cells or *mycetocytes* (Fig. 29) which are sometimes grouped into organs known as *mycetomes*. Micro-organisms of the latter kind are known from many insects and are often transmitted from the mother to her progeny via the eggs; the insects may die rapidly when deprived of their micro-organisms, but their true physiological role has been established only in a few cases.

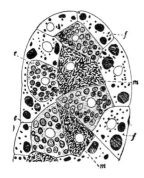

Fig. 29. Lobe of fat body from *Blatta orientalis.* × 650

e, excretory cell with concretions; *f*, fat cell; *m*, mycetocyte containing bacteroids. (*Adapted from* Gier, 1936)

THE RESPIRATORY SYSTEM

In almost all insects respiration takes place by means of internal tubes known as *tracheae*, which convey oxygen directly to the tissues. These tubes ramify over and among the various organs, and in insects with an open respiratory system they communicate with the atmosphere through one or more pairs of respiratory apertures or *spiracles*. Some aquatic and endoparasitic insects have a closed tracheal system. There are no open spiracles and oxygen either diffuses into the tracheae over the greater part of the cuticle or passes in mainly through the integument of special respiratory outgrowths known as *gills* or *branchiae*. In all cases the respiratory organs are derived from the ectoderm; the tracheae are developed as tubular invaginations and the gills arise as outgrowths. Histologically, both types of organ consist of a thin layer of cuticle, an epidermal layer and usually a basement membrane, all of which are continuous with similar layers of the general integument.

The Spiracles. The spiracles are paired openings usually situated on the pleura of the meso- and metathorax and along the sides of the abdomen. In generalized insects such as Orthoptera and in

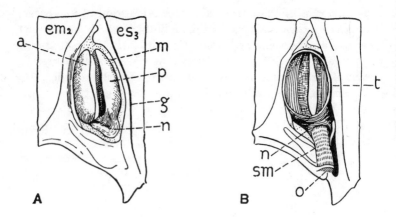

Fig. 30. Metathoracic spiracle of a grasshopper (*Dissosteira*)

A, outer view. B, inner view. em_2, mesepimeron; es_3, metepisternum; g, intersegmental fold; m, membrane; t, trachea. Further explanation in the text. (*Adapted from* Snodgrass, 1929)

some larvae there are ten pairs of spiracles, two pairs being thoracic and eight pairs abdominal in position. Reductions in this number, however, are very frequent. In those cases where prothoracic spiracles occur their presence on this segment is apparently due to migration from behind in the course of evolution. Although spiracles are wanting in most Collembola, a single pair is present on the neck in some Sminthuridae. In the Diplura some Japygidae carry four pairs of thoracic spiracles, two on the mesothorax and two on the metathorax.

Many structural types of spiracles occur; the simplest are found in some Apterygota, where they are merely openings into the tracheal system, surrounded by a simple rim or *peritreme*, but not provided with mechanisms for regulating the size of the aperture. More usually the spiracular opening leads into a chamber, the *atrium*, at the base of which is the tracheal orifice. The passage of air into and out of the tracheae is usually regulated by a *spiracular closing mechanism*, of which two main types occur: (*a*) An external closing apparatus, such as is well shown in the thoracic spiracles of a grasshopper (Fig. 30, 31 A), consists of two movable lip-like sclerites (*a, p*) united by a ventral lobe (*n*). The lips open through their own elasticity but are closed by the contraction of an occlusor muscle (*sm*) which arises from a process (*o*) near the coxal cavity. (*b*) The internal closing mechanism may take many forms. That shown in Fig. 31 (B, C, D) has no external lips but one wall of the atrium is movable (*v*) while the other (*d*) is fixed. The movable wall is prolonged into a process (*q*) to which are attached the occlusor (sm_1) and dilator (sm_2) muscles. Contraction of the former causes the movable wall to close the tracheal aperture while the antagonistic dilator opens it. Spiracles of the second type often have the atrial wall produced into interlacing branched hair-like processes or *trabeculae* forming a filtering apparatus. This device allows the free passage of air, while the entry into the

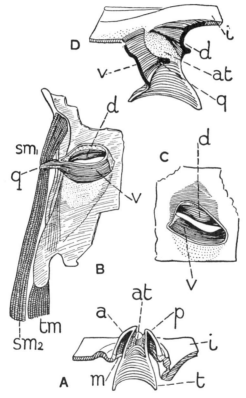

Fig. 31. Spiracles of a grasshopper (*Dissosteira*)

A, D, sections through metathoracic and 1st abdominal spiracles respectively. B, inner view and, C, outer view of 1st abdominal spiracle. *at*, atrium; *i*, integument; *tm*, tympanal muscles. Further explanation in the text. (*Adapted from* Snodgrass, 1929)

atrium of foreign particles or water is prevented. Provision of this kind is common in Lepidopterous larvae. In larvae of the blowfly (*Calliphora*) and other Diptera there is no closing apparatus. The anterior spiracles consist of short lobes perforated at their apices and the posterior spiracles have three openings guarded by trabeculae (Fig. 86 D and E).

It should be noted that the spiracles, in addition to their respiratory function, are the apertures through which the old tracheal linings are drawn out when the insect moults and that they are also a major site of water loss. The latter disadvantage is partially overcome by the closing mechanism or the atrial processes of Dipterous larvae, but these in turn have necessitated the evolution of special devices for withdrawing the tracheal linings at ecdysis.

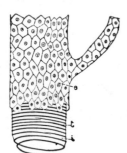

Fig. 32. Portion of a trachea (highly magnified) *e*, epithelial layer (tracheal matrix cells); *i*, cuticular intima with taenidium, *t*

The Tracheae. The tracheae when filled with air present a silvery appearance. They are lined by cuticle continuous with that of the body-wall. This lining has a characteristic striated appearance due to thread-like ridges which run helically around the inner circumference and form the so-called spiral thread or *taenidium*. Continuity of the spiral is often interrupted and a new spiral then begins (Fig. 32). The function of this spiral thickening is to keep the tracheae distended and thereby allow the free passage of air. Externally there is a layer of polygonal cells that secrete the cuticular lining. When a trachea is followed in its branching it finally enters a stellate *end cell* (Fig. 33) and there divides into tracheal capillaries or *tracheoles*. These are less than 1 μm in diameter and their taenidia are so delicate as to be visible only under the electron microscope. The tracheoles end in the tissues in various ways. In the gut and salivary glands they ramify and pass between the cells

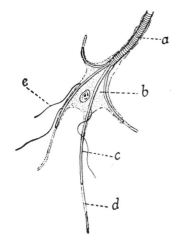

Fig. 33. Tracheal end cell in head of mosquito larva *Aedes aegypti*

a, trachea; b, tracheal end cell; c, tracheole containing air; d, terminal part of tracheole containing fluid; e, fine branch of tracheole. (*After* Wigglesworth)

without penetrating them. In the fat-body and rectal papillae, however, they may enter the cells, while in the flight muscles there is a network of intracellular tracheoles.

In *Campodea* and some other Apterygotes the tracheae arising from each spiracle remain unconnected with those from the others. Elsewhere a more efficient system of longitudinal and transverse (segmental) connecting tracheae has been evolved. In a typical segment (Fig. 34) three principal tracheae arise from the main longi-

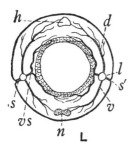
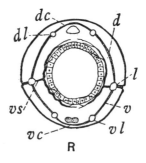

Fig. 34. Diagrammatic transverse sections of the abdomen showing two types of tracheal system

L, with lateral longitudinal trunks, *l*: R, with dorsal, *dl*, and ventral, *vl*, longitudinal trunks, together with dorsal, *dc*, and ventral, *vc*, commissural trachea. *d*, *vs*, *v*, dorsal, visceral and ventral tracheae; *s*, spiracular trachea; *h*, heart; *n*, ventral nerve cord

tudinal trunk (*l*) near the point where the spiracular trachea (*s*) joins it. These tracheae are: (i) a *dorsal trachea* (*d*), supplying the dorsal muscles and heart; (ii) a *visceral trachea* (*vs*) passing to the gut, fat-body and gonads, and (iii) a *ventral trachea* (*v*) which supplies the ventral musculature, nerve cord and, in the thorax, the legs and wings. Secondary longitudinal trunks (dorsal, visceral and ventral) may also be developed (Fig. 34 R).

The Air-sacs. In many insects thin-walled sac-like dilations or *air-sacs* occur in large numbers. The walls of these usually are extremely delicate in structure and devoid of any special thickening (Fig. 35). In the cockchafter (*Melolontha*) and in grasshoppers (Acrididae) air-sacs are present in large numbers as dilations of the smaller tracheae. In the housefly (*Musca*) and in many other Diptera and in bees the main tracheal trunks are dilated to form extensive air-sacs, especially in the abdomen. Air-sacs are mainly developed in swiftly flying insects, and one of their functions is to provide increased ventilation for the tracheal system. They

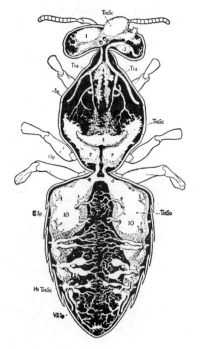

Fig. 35. Tracheal system of worker honey-bee seen from above

(One pair of abdominal air-sacs removed and transverse ventral commissures of abdomen not shown.) The air-sacs (TraSc) are indicated in arabic numerals; *sp*, spiracles. (*After* Snodgrass, U.S. *Bur. Entom. Tech. Ser.* No. 18)

respond very rapidly to the increase and decrease of pressure resulting from respiratory movements and consequently greatly augment the volume of air inspired and expired with each movement. Periodic compression of the thoracic air-sacs also enables oxygen to be pumped into the actively respiring flight muscles.

Types of Tracheal System. Several types of tracheal system are recognized, depending on the number and position of the open, functional, spiracles (Fig. 36). The *holopneustic* type of system has all 10 pairs of spiracles open, the *peripneustic* type has all but a few open, and the *amphipneustic* type has only the first and last

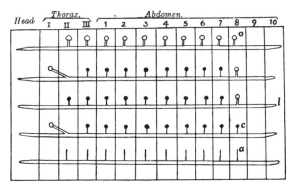

Fig. 36. Types of tracheal system
a, spiracle absent; *c*, spiracle closed; *o*, functional spiracle; *l*, longitudinal tracheal trunk

pairs open. In *propneustic* and *metapneustic* systems the first or last pairs respectively are the only open ones, while the *apneustic* type has no functional spiracles and respiration occurs through the general body surface or gills. The tracheal systems with few or no open spiracles are found in aquatic and endoparasitic insects. In all the foregoing types, the non-functional spiracles are closed or reduced to minute scars and each is connected to the longitudinal tracheae by a solid *stigmatic cord*. The latter helps to anchor the tracheal system in position and serves to pull out the old tracheal linings when the insect moults. If, in the ensuing instar, the non-functional spiracle is to be replaced by an open one, the new spiracular trachea – which forms around the stigmatic cord

– remains tubular; otherwise it shrivels to form a new stigmatic cord. In some insects one or more pairs of spiracles have disappeared completely and the tracheal system may lose much of its originally metameric nature. Such systems may be referred to as *hypopneustic* (e.g., some Coccoidea).

Respiration. In the smaller or less active terrestrial insects oxygen passes along the tracheal system, from the spiracles to the finer tracheoles, by the process of gaseous diffusion. This is possible because of the difference in partial pressure of oxygen between the atmosphere and the tracheolar endings, where the gas is constantly being removed by the respiring tissues. Surprising though it seems, a difference in partial pressure of only 2–3 mm. of mercury is adequate to supply by diffusion all the oxygen that some insects require. A similar process, in the reverse direction, would also account for the removal of carbon dioxide. The latter, however, diffuses through insect tissues much more readily than does oxygen, so it is likely that much of it – perhaps 25% – is eliminated through the tracheal walls and the cuticle of the body surface rather than through the spiracles. The walls of the tracheoles are more permeable to oxygen than are the other parts of the tracheal system and it is therefore through these fine branches that the tissues receive most of their oxygen supply. There is usually some liquid in the endings of the tracheoles, but in active muscular tissue the rise in osmotic pressure which accompanies contraction results in the absorption of this liquid. Air can then penetrate further along the tracheoles and so increase the rate at which oxygen will diffuse into those tissues which are in great need of it (Fig. 37).

The physical law governing the diffusion of oxygen along the tracheal system is such that large or very active insects would not receive enough by this means alone. Diffusion is therefore supplemented in these insects by ventilation of the tracheal system through respiratory movements. Expiration is effected by various muscles of the abdomen, whose contraction flattens the body in Orthoptera and Coleoptera, while among Hymenoptera and Diptera telescopic movements of the abdominal segments result. Inspiration is usually effected by the elasticity of the body segments as they regain their original shape. Most of the tracheae are circular in cross-section and therefore only compressible with

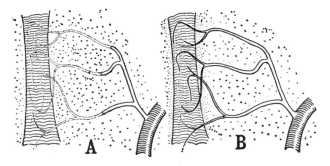

Fig. 37. Tracheoles running to a muscle-fibre: semi-schematic
A, muscle at rest; terminal parts of tracheoles (shown dotted) contain fluid. B,
muscle fatigued: air extends far into tracheoles. (*After* Wigglesworth)

difficulty, but they are readily extensible in the direction of their long axis. The main tracheal trunks are often oval in cross-section and are readily compressed while the air-sacs are even more prone to collapse. By alternately compressing and dilating the main tracheae and air-sacs, respiratory movements movements therefore bring about mechanical ventilation of the larger tubes. In some insects it is known that there is a directed air flow in the tracheal system. Among the Acrididae in particular it has been shown that the first four pairs of spiracles can remain open while the remaining six pairs are closed and *vice versa*, with the result that the former serve for inspiration and the latter for expiration. In larvae of the beetle *Dytiscus* it has been estimated that the tracheal system is emptied during strong expiration of nearly two-thirds (64 mm^3) of its total capacity (107 mm^3), the remainder being changed by diffusion.

Respiratory activity is regulated by the opening or closing of the spiracles (diffusion control) and, in those insects with ventilation, by variations in the intensity or frequency of respiratory movements (ventilation control). In each segment of the body, the muscular contractions responsible for these changes are governed by a primary respiratory centre situated in the corresponding ganglion of the ventral nerve cord. Further co-ordination and control is exercised by the so-called secondary respiratory centres of the brain (diffusion control) or thorax (ventilation control). Lack of oxygen or accumulation of carbon dioxide stimulates these centres and results in the opening of the spiracles or the initiation of ventilatory movements.

Because the tracheal system conveys oxygen into or close to the respiring tissues the blood of insects plays only a small part in respiration, and its oxygen capacity is apparently no greater than can be accounted for by simple solution. The only known exceptions to this are the aquatic larvae of some Chironomid midges, known as blood-worms, whose plasma contains haemoglobin in solution. The haemoglobin, however, is such that it only transfers oxygen to the tissues when the partial pressure of oxygen is very low. It therefore appears to be an adaptation enabling the larvae to survive in the poorly oxygenated conditions which they sometimes have to endure.

Respiration in Aquatic Insects. Some aquatic insects have an open tracheal system and obtain oxygen from the atmosphere through one or more pairs of functional spiracles. In others the tracheal system is closed and oxygen diffuses into it from the water through part or all of the integument. Those forms with an open tracheal system show varying degrees of specialization. In the larvae of some Culicidae and Syrphidae, for example, the posterior spiracles are situated on siphon-like processes which penetrate the surface film and so make contact with the atmosphere. Again, in larvae and pupae of a few species of Coleoptera and Diptera (e.g., *Donacia* and *Taeniorhynchus*) the spiracles are placed on pointed processes which can be inserted into the air-containing cavities of submerged aquatic plants. Many Coleoptera and Hemiptera (e.g., *Dytiscus* and *Notonecta*) have aquatic adults which, when they submerge, carry air-bubbles trapped beneath the elytra or on other parts of the body. These air stores are in contact with open spiracles and oxygen is drawn from them. They also act as 'physical gills' which extract dissolved oxygen from the water. This function depends on the invasion coefficient of oxygen between water and air being three times that of nitrogen. It follows that as the partial pressure of oxygen in the bubble falls, further oxygen diffuses into it from the water faster than the nitrogen of the bubble diffuses out. In this way far more oxygen can be extracted from the water than was originally present in the bubble. Ultimately, however, the slow dispersal of nitrogen causes the bubble to grow smaller and it must therefore be renewed repeatedly by the insect ascending to the surface. A few insects, however, like the Hemipteran *Aphelocheirus*, have evolved a method of 'plastron respiration' in which a thin film of air – the plastron – is held against the body by special hairs or other cuticular processes in such a way that it cannot decrease in volume. It therefore functions as a permanent physical gill and such insects, although they have an open tracheal system, can remain submerged throughout life, extracting all their oxygen from the water.

In some insects with a closed tracheal system the integument is very thin and overlies a network of fine tracheal branches. Oxygen then diffuses in from the water over most of the body surface, e.g., larvae of *Chironomus* and *Simulium*. In other species, however, the body wall is produced locally into *tracheal gills* – thin-walled outgrowths which are well supplied with tracheae – and much of the respiratory exchange is restricted to these organs. They may be filamentous or lamellate and occur on various parts of the body, e.g., the abdomen in nymphs of mayflies and Zygopteran dragonflies (Figs. 65 and 67) and the thorax of some stonefly nymphs. The body wall adjacent to a spiracles may be produced into a so-called *spiracular gill*; insects so equipped can respire in air or under water. In Anisopteran dragonfly nymphs the gills occur on the inner walls of the hind gut – the so-called rectal gills – and water is repeatedly drawn in and expelled through the anus. A few aquatic insects, such as some Dipterous larvae, possess structures known as *blood gills* or anal papillae. These are small tubular outgrowths of the body, usually near the anus, with few or no tracheae. They were formerly regarded as respiratory organs but are now thought to be more concerned with absorbing inorganic salts from the water and so regulating the ionic composition and osmotic relations of the blood.

Respiration in Endoparasitic Insects. These live immersed in the body fluids of their host and therefore show respiratory adaptations which resemble those of aquatic insects. In some, such as the larvae of Tachinid flies, there is an open metapneustic respiratory system and the parasite perforates the integument of its host or one of the tracheal trunks. By inserting the spiracular region of the body into the opening thus formed it can draw on a supply of atmospheric oxygen. In others, such as the early larval instars of many parasitic Hymenoptera, the tracheal system is absent or

filled with liquid and oxygen diffuses into the blood from the body
fluids of the host. In the older larval instars a closed tracheal system is present and oxygen diffuses through the integument, which
is provided with a network of subcutaneous tracheae. Some endoparasitic larvae bear tail-like or vesicular outgrowths which are
known in a few species to act like gills; in a few others, however,
they are not specially concerned with respiration, and in most cases
their physiology has not been studied.

THE CIRCULATORY SYSTEM AND ASSOCIATED TISSUES

The body cavity in insects is a haemocoele which contains the
circulating blood. All the organs and tissues are bathed with this
liquid and perform their functions through exchanges with it. The
haemocoele in the majority of insects is divided into sinuses by

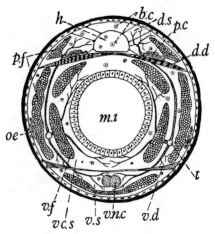

Fig. 38. Schematic section
across the abdomen of an
insect showing circulatory
system and fat-body

b.c, blood cells; *d.d*, dorsal
diaphragm; *d.s*, dorsal sinus; *h*,
heart; *m.i*, mid intestine; *oe*,
oenocytes; *p.c*, pericardial
cells; *p.f*, parietal fat-body; *t*,
main trachea; *vc.s*, visceral
sinus; *v.d*, ventral diaphragm;
v.f, visceral fat-body; *v.n.c*,
ventral nerve-cord; *v.s*, ventral
sinus

fibro-muscular septa or diaphragms (Fig. 38). The *dorsal diaphragm* is the septum most constantly present; it extends across
the abdominal cavity just above the alimentary canal and in this
way divides the haemocoele into a *dorsal* or *pericardial sinus*, containing the dorsal vessel, and a very large *visceral sinus* representing
the remainder of the body cavity. In some insects there is also a
ventral diaphragm stretching across the abdominal cavity above the
nerve cord and thus demarcating a *ventral* or *perineural sinus*. Pairs

of *aliform muscles* arise from the abdominal terga and spread out fanwise over the dorsal diaphragm.

The **dorsal vessel** (Fig. 39) is the main conducting organ of the circulatory system and is divided into the heart and the aorta. The *heart* is a muscular contractile tube situated in the median line of the pericardial sinus just above the dorsal diaphragm. It is held in position by fibrous strands connected to the body wall and the diaphragm. As a rule, the heart is a narrow continuous vessel whose sides are perforated with vertical slit-like openings or *ostia*. The margins of the ostia may be prolonged inward to form valves which prevent the return of blood from the heart into the peri-

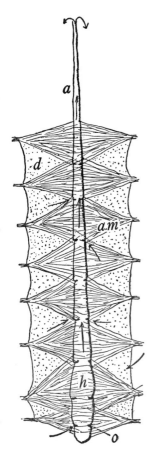

Fig. 39. Dorsal vessel of a beetle – ventral side

a, aorta; *a.m*, aliform muscle; *d*, dorsal diaphragm; *h*, heart; *o*, ostium. Arrows show course of circulation

cardial sinus. In other cases the heart shows a series of dilations or chambers usually corresponding in number to the pairs of ostia and of aliform muscles. While there may be a chamber of the heart to each segment of the abdomen and to the second and third segments of the thorax, as in cockroaches, the number of chambers is generally much fewer and may even be reduced to a single terminal enlargement. The *aorta* is the slender anterior prolongation of the dorsal vessel which carries the blood into the head where it opens behind or beneath the brain. The wall of the heart is muscular and is composed of flattened cells whose outer cytoplasm is differentiated into striated muscle fibrils. These cells are bounded externally and internally by a delicate membrane which may be regarded as a sarcolemma. Apart from the aorta there are few closed vessels associated with the circulatory system. Among the best known are the abdominal arteries of cockroaches and the antennal arteries. In *Blatta* and the honey bee the blood is propelled through the antennal arteries by special *pulsatile organs* (accessory hearts) situated at their bases. Pulsatile organs are also found in the thorax where they maintain circulation of the blood in the wing veins.

The **blood** or *haemolymph* is the only extracellular fluid in insects. It is clear, colourless or very pale yellow or green, and consists of the liquid *plasma*, in which are suspended numerous colourless blood cells or *haemocytes*. The plasma, consisting of about 85% water, usually has a pH of 6·4–6·8 and contains amino acids, proteins, fats, sugars (mainly the disaccharide a-trehalose) and inorganic salts (mainly sodium, potassium and chloride ions). It serves as a store of water, transports food materials and hormones and conveys limited quantities of oxygen and carbon dioxide (p. 86). Its hydrostatic properties enable it to transmit pressure changes, as in the splitting of the old cuticle at moulting or the eversion of the wings of newly emerged adults. The haemocytes (Fig. 40) are very variable in form and many different types have been distinguished, though some of these are probably developmental stages or the result of temporary changes in shape. The circulating blood contains 1000–100 000 cells per mm^3 but many more are found adhering to the surfaces of the viscera, where they sometimes form more or less fixed phagocytic organs. There are four main categories of haemocytes: (1) *prohaemocytes*, small cells

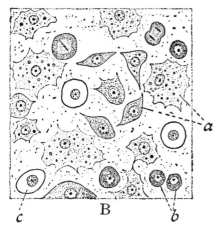

Fig. 40. Haemocytes
a, phagocytes; *b*, proleucocytes; *c*,
oenocytoid. (*From* Wigglesworth)

with deeply staining cytoplasm and large nuclei; they are often seen undergoing mitosis and are regarded as young forms of haemocytes. (2) *Phagocytes*, which are basiphil and of variable appearance and character; their amoeboid activities enable them to digest tissue debris and invading bacteria, and they greatly increase in numbers during ecdysis and metamorphosis. Three kinds have been distinguished: plasmatocytes, granular cells and spherule cells. The phagocytes have the property of congregating around and enclosing foreign bodies; they also collect at the site of a wound, forming a plug which facilitates healing. (3) *Oenocytoids*, which are rounded cells with acidophil cytoplasm. They bear a resemblance to small oenocytes, but their function is not known. (4) *Cystocytes*, which are rounded cells with relatively large nuclei and are probably involved in the clotting of the blood.

The **circulation** of the blood begins in an anterior stream that is maintained by waves of contraction, passing from behind forward, over the heart. During diastole blood is drawn into the heart, through the ostia, under a negative pressure. During systole a positive pressure is set up and the blood is driven forward in the heart cavity and eventually leaves the open anterior end of the aorta in the head. Here some of it circulates through the antennae, but ultimately the blood enters the visceral sinus after a proportion has circulated through the legs and wing veins. Undulatory movements of the ventral diaphragm direct a blood-flow backward in

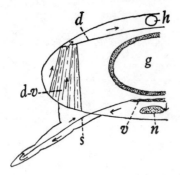

Fig. 41. Cross-section of thorax of *Blatta* showing course of circulation *d*, dorsal diaphragm; *d–v*, dorsoventral muscle; *h*, heart; *n*, nerve cord; *s*, septum; *v*, ventral diaphragm. (*Adapted from* Brocher)

the perineural sinus. Leaving the latter, through spaces along its sides and posterior end, the blood ascends among the viscera. It then becomes drawn into the pericardial sinus by contractions of the aliform muscles. This alters the contour of the dorsal diaphragm and passage of the blood into the sinus occurs through perforations in this membrane (Fig. 41). The property of rhythmical contraction of the heart lies in the muscle fibrils of its walls. This automatism is evident in the isolated heart or even in severed portions of it since they continue to beat in the usual rhythmic manner. The pulsation rate is influenced by many external, endocrine and pharmacological factors. Thus, it is increased as the temperature is raised; it may vary in different instars of the same insect. It has long been known that in *Sphinx ligustri* the heart beat of the moth is at the rate of 40–50 per minute when at rest and 110–140 during activity. In the larva the rate is highest (80–90 per minute) in the early instars and lowest in the pupa when it declines to 10–12 beats per minute; during hibernation it ceases almost entirely.

THE EXCRETORY ORGANS, FAT-BODY AND OTHER HAEMOCOELIC STRUCTURES

The function of an excretory system is to maintain a relatively constant internal environment for the tissues of the body. Among other things, this involves the removal of the nitrogenous products of protein breakdown and the regulation of the ionic composition of the haemolymph. The chief excretory organs are the Malpighian tubules, whose structure is described on p. 75 and which work in conjunction with the hind gut. Some other tissues are, or were, thought to play a subsidiary role and are therefore also conveniently discussed here.

Malpighian Tubules. These remove excretory materials from the blood in the form of urine, which is secreted into the lumen of the tubule and ultimately discharged into the hind gut, where its composition may be modified by resorption before it passes out with the faeces. The urine may be a clear aqueous liquid or a thick suspension and its composition differs in different insects. The principal nitrogenous material is uric acid – or perhaps its ammonium, sodium or potassium salts – usually appearing in the form of crystalline spheres. Some species secrete substantial quantities of allantoin, allantoic acid, urea or ammonia. Inorganic salts, in solution or as granules or spheroidal concretions, also occur.

Various functional modifications of the Malpighian tubules occur in different insects (Fig. 42); many of these permit the elimination of nitrogenous materials while conserving the limited supply of water available to terrestrial organisms. In *Rhodnius* (Fig. 42 A) the distal part of each tubule secretes a solution of sodium or potassium acid urate while the proximal part reabsorbs much of the water and bases, the latter in the form of bicarbonate. Uric acid is thus precipitated as crystalline spheres in this part of the tubule while the water and bases are recirculated continually. Further absorption of water and sodium ions occurs in the rectum. In *Lepisma*, Orthoptera, Neuroptera and many Coleoptera (Fig. 42 B) the Malpighian tubules only contain fluid while precipitation of a white crystalline mass of uric acid takes place in the rectum through whose walls water absorption is carried out. In mosquitoes and Muscid flies (Fig. 42 C) the Malpighian tubules contain solid uric acid throughout their length and a method of precipitation occurs different from that found in *Rhodnius*. In many Coleoptera and the larvae of Lepidoptera the distal parts of the Malpighian tubules are closely attached to the walls of the rectum. This device seems to facilitate water conservation by using the combined absorptive capacity of the rectum and Malpighian tubules. In the mealworm (*Tenebrio*) (Fig. 42 D), where such an arrangement occurs, the Malpighian tubules only contain clear fluid and the rectum is mainly occupied by a dry mass of uric acid, apparently precipitated by the almost complete absorption of the available water.

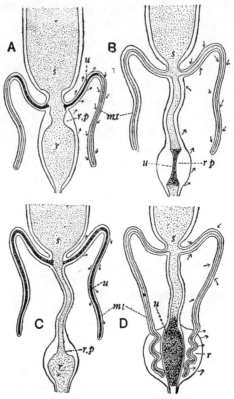

Fig. 42. Diagrams of different types of excretory system in insects (*Adapted from* Wigglesworth. Explanation in the text.) *mt*, Malpighian tubules; *r*, rectum; *rp*, rectal papillae; *s*, mid intestine; *u*, zone of precipitation of uric acid. The arrows indicate presumed course of circulation of water and base

Other excretory products include calcium salts that are sometimes taken into the body in quantities greatly above requirements. They are present in the Malpighian tubules either as amorphous granules or less frequently as solid spheres or crystals. The most general compound is calcium carbonate, which is usually stored during larval life and often used in various ways during metamorphosis, disappearing by the time the imago is reached. Many Dipterous larvae contain calcium carbonate either in the Malpighian tubules or in special cells of the fat body; in phytophagous larvae of the family Agromyzidae the calcium occurs as laminated bodies or 'calcosphaerites' that are seen clearly in those of the celery fly (*Euleia heraclei*) and other species. At meta-morphosis the calcium carbonate dissolves in the blood and becomes deposited on the inner wall of the puparium. Many larvae of the Cerambycidae (p. 181) line the pupal burrow with lime and also close it with an operculum composed of similar material. In the Phasmidae the chorion of the eggs is hardened by becoming impregnated with calcium salts.

The physiology of excretion is unusual in the more highly adapted aquatic insects. Ammonia may be the principal excretory product and there are no special mechanisms for conserving water. On the other hand there may be very active reabsorption of in-organic ions in the hind gut.

The Fat-body (Fig. 38) is present in all insects and is derived from the mesoderm of the walls of the embryonic coelomic cavities. It sometimes shows a segmental disposition and occurs as loose strands, sheets and lobes of tissue. Generally there is a visceral layer around the gut and a peripheral layer beneath the integument. Since the fat-body lies in the haemocoele it is im-mersed in blood which also circulates through its interstices. The fat-body is therefore well adapted for its main function – the synthesis and storage of reserve materials. These include fat, pro-teins and glycogen, all of which can be mobilized as required. In the newly-hatched insect the cells of the fat-body are rounded, with a homogeneous cytoplasm free from vacuoles or inclusions. During growth these cells increase in size, become vacuolated and their boundaries are then hard to see; nuclear changes also occur. In a starved insect the reserve food materials mentioned may become used up, but normally they are drawn upon at certain periods only, i.e., during the change from larva to pupa, during hibernation, and to some extent at each ecdysis. In adult insects the fat-body is often more developed in the female, where it pro-vides nutriment for egg development.

The fat-body also performs an excretory function. In Collem-bola, Hymenopterous larvae, Orthoptera, etc., special *excretory cells* containing deposits of urates are present among the ordinary cells of the tissue (Fig. 29). The excretory cells serve in the main as storage cells until their products are eliminated at the time of pupation. In Collembola, which lack Malpighian tubes, urate con-cretions are deposited and increase in size throughout life; much the same is stated to occur in *Lepisma*, the Dermaptera and Orthoptera, where the Malpighian tubes apparently eliminate little uric acid.

Nephrocytes. These are special cells which occur scattered in the haemocoele or concentrated in certain restricted regions. They are commonly found in strands along each side of the heart, where they are known as pericardial cells (Fig. 38) and are often binucleate. In the larvae of Cyclorrhaphan Diptera they are arranged in a garland-like structure between the salivary glands. Their distinctive feature is a capacity for accumulating the dyestuff ammonia carmine when it is injected into the haemocoele. For this reason they were formerly regarded as organs of 'storage excretion'. This is now thought unlikely, though under experimental conditions they can accumulate a wide variety of colloidal particles by pinocytosis and thus resemble the reticulo-endothelial cells of vertebrates.

Oenocytes. These are usually large cells often of a wine-yellow colour to which they owe their name. They occur in larvae and adults of all orders and rank among the largest cells of the body, sometimes measuring up to about 180 μm across. Their large nuclei, and dense eosinophil cytoplasm with an external limiting membrane, are characteristic features. They arise from the ectoderm of the embryo as segmental groups of cells situated close behind the invaginations that give rise to the abdominal spiracles (Fig. 38). They may remain associated with the epidermis as in *Blatta* or migrate into the peripheral fat-body (*Locusta, Anopheles*) or come to lie in close association with spiracular tracheae (Lepidoptera). Oenocytes show a cycle of morphological changes and secretory activity at the time of moulting, usually increasing greatly in size just before that process; the cytoplasm becomes vacuolated and the cells may become lobed as in *Rhodnius*. It seems probable that they elaborate some of the materials used in the construction of the cuticle; whether they have other functions is not clear.

THE GLANDS OR ORGANS OF SECRETION

Two main types of special secretory organs occur in insects; (i) the *exocrine glands*, which are provided with ducts and discharge their secretions outside the body or into the lumen of one or other of the viscera; and (ii) the *endocrine glands*, which have no ducts and whose secretions, known as *hormones*, usually diffuse into the blood, which transports them to all parts of the body.

Exocrine Glands. Only a few of the more important types of exocrine glands may be mentioned here. *Wax glands* are well seen on the abdominal sterna of worker honey bees and are also common in such Homoptera as the scale-insects (Coccoidea) and some aphids. They comprise one or more epidermal cells, often discharging through plate-like pores, or aggregations of pores, in the cuticle. Associated with the mouthparts of all insects are one or more paired secretory organs, the most important of which are the *labial glands*. Their tubular ducts unite and discharge by a common canal near the base of the hypopharynx, and since they normally secrete saliva they are commonly known as the salivary glands. They may be tubular, lobulate or diffuse structures, sometimes provided with a reservoir. In the larvae of Lepidoptera, sawflies and Trichoptera the labial glands secrete silk and saliva is produced by the mandibular glands. Some insects are provided with *repugnatorial glands*, such as occur on the abdominal terga of many immature Heteroptera, on the metapleura of their adults, and near the anus in many Coleoptera. Their function is probably defensive and this is also true of the various types of *poison glands*, e.g., those associated with the setae of some Lepidopteran larvae or with the sting of wasps and bees (p. 40). *Attractant glands*, producing volatile secretions attractive to the opposite sex, commonly occur in male Lepidoptera – often on the wings, where they are associated with special scales, the *androconia*. In some female Lepidoptera the secretion of glands near the end of the abdomen can attract males from considerable distances. A variety of other pheromone-secreting glands occur in different insect groups.

Endocrine Glands. Various small, paired glands in the anterior part of the body produce hormones which play a very important role in the control of moulting and metamorphosis (pp. 119, 129) and may also exert other physiological effects. Of these glands, the most important are: (i) The *neurosecretory cells* of the brain. These are groups of modified nerve-cells, situated in the dorsal part of

the protocerebrum that produce a peptide hormone which activates the thoracic glands (thoracotropic hormone). Another neurosecretory hormone known as bursicon, which controls the
hardening and darkening of the cuticle of newly moulted insects,
also seems to be secreted by the brain, though it is discharged
from the thoracic or abdominal ganglia. Neurosecretory cells also
occur scattered or in small groups among the ganglia of the ventral
nerve cord; their function is not well understood, but in *Rhodnius*
they produce a diuretic hormone. (ii) The *corpora cardiaca*. These
are usually a pair of small bodies lying one on each side of the
aorta immediately behind the brain, to which they are connected
by two pairs of nerves. They include the swollen endings of neurosecretory axons from the pars intercerebralis cells and serve as
so-called neurohaemal organs, responsible for discharging into the
haemolymph the neurosecretory materials produced by the brain.
In addition, the corpora cardiaca include intrinsic neurosecretory
cells, sometimes segregated into a distinct glandular lobe. Extracts
of the corpora cardiaca may therefore include brain hormones as
well as substances secreted by the intrinsic cells. Such extracts are
known in some cases to affect the lipid and carbohydrate levels
of the blood and to control water balance; they may also include
materials whose effects are pharmacological rather than associated
with the normal physiology of the insect. (iii) The *thoracic glands*.
These occur in the prothorax, though apparently homologous
organs with a similar function are found in the head of some
lower Exopterygotes. They produce hormones (ecdysone and its
derivatives) which induce the immature insect to moult and they
degenerate in the adult (except in the Thysanura where moulting
continues after sexual maturity). (iv) The *corpora allata*. These
glands are closely associated with the stomatogastric nervous system (p. 55; Fig. 18), and produce the so-called juvenile hormones.
These tend to inhibit the appearance of adult characteristics in the
developing insect and while they are being produced in sufficient
quantity they ensure that the moults induced by the thoracic gland
hormone give rise to the normal sequence of nymphal or larval
instars. Towards the end of juvenile life, however, the corpora
allata become much less active and moulting is then accompanied
by the more or less abrupt development of adult features, such
as is strikingly shown in the metamorphosis of holometabolous
insects (p. 121 ff). The corpora allata resume activity in the adult,

when their secretions may be necessary for the full development of the ovaries and the accessory reproductive glands of both sexes.

THE REPRODUCTIVE SYSTEM AND REPRODUCTION

The reproductive organs, which differ appreciably in the two sexes, consist of (i) a pair of gonads, which are derived from mesoderm; (ii) the system of efferent ducts, which is usually partly mesodermal and partly ectodermal, and (iii) the various annexes, such as accessory glands and structures for the temporary reten-

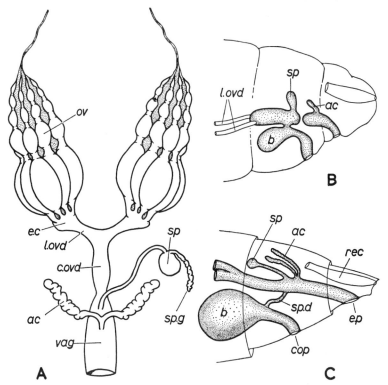

Fig. 43. Female reproductive system

A. In a typical insect. B. Development of efferent ducts in pupa of a higher Lepidopteran. C. Adult efferent ducts in a higher Lepidopteran. *ac*, accessory gland; *b*, bursa copulatrix; *cop*, copulatory aperture; *c.ovd*, common oviduct; *e.c*, egg calyx; *e.p*, egg pore; *l.ovd*, lateral oviduct; *ov*, ovary; *rec*, rectum; *sp*, spermatheca; *sp.d*, sperm duct; *sp.g*, spermathecal gland; *vag*, vagina

tion of the spermatozoa. Closely associated with the external openings of the reproductive system are the external genitalia, discussed on p. 37.

The Female Reproductive System

The female gonads or *ovaries* discharge the eggs into a pair of tubular *lateral oviducts*, almost always formed from mesoderm. In the mayflies, each oviduct opens by a separate gonopore behind the seventh abdominal sternum and a condition recalling this primitive state of affairs is passed through in the immature stages of most insects. In the adults, however, the lateral oviducts generally run into an unpaired *common oviduct* which is then continued into a wider passage or *vagina*, whose orifice lies behind the eighth or ninth abdominal sternum (Fig. 43 A). These unpaired parts of the reproductive system develop from one or two ectodermal invaginations. Further ectodermal ingrowths give rise to the *spermatheca* or *receptaculum seminis* and to a pair of accessory glands; in the adult these organs usually open by short ducts into the common oviduct or vagina.

The Ovaries. Each ovary usually consists of several egg tubes or *ovarioles*, which are only occasionally bound together by an outer membrane into a more or less compact organ. The ovarioles may open into the lateral oviduct one behind the other or they are arranged in a radiating manner and all enter it at about the same place. There are commonly 4–8 ovarioles in each ovary, but some Hymenoptera may have more than 200 and even this number is exceeded in the queens of some termites. Each ovary of some viviparous Diptera, on the other hand, consists of a single ovariole. A typical ovariole (Fig. 44) consists of a *terminal filament*, a *germarium*, and a *vitellarium*. The terminal filaments of all the ovarioles of one side are usually united to form a *suspensory ligament*. The apical germarium contains the primordial germ-cells or *oogonia*; these later become differentiated into *oocytes* and also, when they are present, into nurse cells or *trophocytes*. The vitellarium consists of a longitudinal series of developing eggs, the smallest and youngest being those nearest to the germarium. As they grow by the deposition of yolk, the eggs distend the ovariole into a series of *follicles* or egg chambers; each egg is enclosed in a layer of follicular epithelium which eventually secretes the *chorion* or egg-shell. In the ovarioles of some insects only the lowermost (i.e., oldest) eggs are completely developed and ready to be discharged (Fig. 44); in others most of the eggs have completed their development by the time oviposition begins.

There are three types of ovarioles (Fig. 44): (i) The *panoistic* type has no nurse cells; the developing egg cell may itself synthesize some components of the yolk and receives others from the

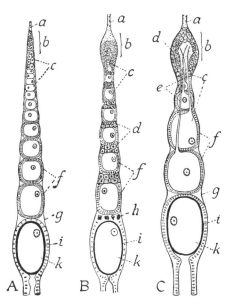

Fig. 44. The three chief types of ovarioles diagrammatic
A, panoistic type. B, polytrophic type. C, acrotrophic type. *a*, terminal filament; *b*, germarium; *c*, oocytes; *d*, nurse cells; *e*, nutritive cords; *f*, follicular cells; *g*, membranous coat; *h*, degenerated nurse cells; *i*, chorion; *k*, egg. (*After* Weber)

undifferentiated follicular epithelium, which transfers materials obtained from the blood. This primitive type of ovariole occurs in the Apterygota, Odonata, Orthoptera and other orders. (ii) In the *polytrophic* type each developing egg is associated with an adjacent group of nurse cells which are responsible for producing some fractions of the yolk. This type is found in most Endopterygota. (iii) In the *acrotrophic* type the nurse cells are confined to the germarium and are connected with the developing oocytes by progressively lengthening cytoplasmic strands, along which some of the nutritive materials pass. This kind of ovariole occurs in the Hemiptera and some Coleoptera.

Genital Ducts and associated structures. Some account of the different kinds of efferent ducts has already been given on p. 100. The vagina commonly ends in a genital cavity whose floor is the *subgenital plate* formed from the seventh or eighth sternum. The spermatheca is usually a sac-like organ which opens into the common oviduct or vagina by a more or less elongate spermathecal duct. The spermatozoa received in mating are stored here and pass down the duct to fertilize the eggs as the latter move along the common oviduct before being laid. The female accessory glands (*colleterial glands*) usually open into the vagina and often secrete an adhesive substance for cementing the eggs to each other or to the substrate on which they are laid. In the cockroaches and praying mantids their secretions produce the sclerotized ootheca or egg capsule in which the eggs of these insects are enclosed.

In the higher Lepidoptera (Fig. 43 C) there are two reproductive openings. The anterior one, on the eighth sternum, is the copulatory aperture. It leads into a large *bursa copulatrix*, which is connected by a narrow *sperm duct* with the common oviduct. The latter is continued into the vagina, along which eggs pass to be discharged through the egg pore on the ninth sternum. The spermatheca joins the common oviduct by a duct in whose wall is a fine *fertilization canal*. Spermatozoa, enclosed in a proteinaceous sac or *spermatophore*, are deposited in the bursa copulatrix, from which they eventually make their way to the spermatheca.

The Male Reproductive System

The mesodermal parts of the male reproductive system are a pair of gonads or *testes* and two lateral ducts or *vasa deferentia*. The latter join a median ectodermal passage, the *ductus ejaculatorius*, which usually opens to the exterior on the *aedeagus* or *penis* (p. 38). In addition to these essential parts there is frequently a pair of *vesiculae seminales* or sperm reservoirs, formed by the enlargement of a part of each vas deferens. *Accessory glands* of ectodermal origin are also commonly present (Fig. 45).

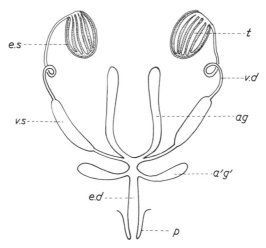

Fig. 45. Diagram of male reproductive system

ag, mesodermal accessory gland; *a'g'*, ectodermal accessory gland; *e.d*, ejaculatory duct; *e.s*, epithelial sheath of testis; *p*, penis; *t*, testis; *v.d*, vas deferens; *v.s*, vesicula seminalis

The Testes. Each testis is composed of tubules or *follicles*, variable in number, which open by narrow passages or *vasa efferentia* into the vas deferens of their side. The testis is covered outwardly by an epithelial sheath often, though inaccurately, known as the peritoneal layer. Each follicle is lined by epithelium resting on a basement membrane, and it is from the cells of this lining that the primordial germ cells are derived. A succession of zones, in which the germ cells are in different stages of development, are to be distinguished. At the apex of a follicle is the *germarium* which comprises spermatogonia among numerous somatic cells. Lower down each spermatogonium becomes surrounded by somatic cells to form a cyst. By the repeated division of a spermatogonium from 64 to 256 *spermatocytes* are produced. In the next zone, or *zone of maturation*, the spermatocytes undergo reduction division so that their chromosome number is halved; each spermatocyte ultimately produces four *spermatids*. There follows a *zone of transformation* in which the spermatids, still enclosed in the cyst wall, are converted into *spermatozoa*; the latter break out of the cyst by lashing movements of their flagella. At first the spermatozoa adhere by their heads in bundles, but they ultimately become free.

Genital Ducts and accessory structures. In mayflies the vasa deferentia remain separate and each enters the penis of its side. From this generalized condition is derived the typical system in which a median ectodermal ingrowth gives rise to the ductus ejaculatorius. Where the vasa deferentia join the anterior extremity of this canal their ends become enlarged ampullae (Fig. 46 A) which unite to form a mesodermal vesicle. The *accessory glands* are of two kinds. In *Blatta* and most Orthoptera numerous glands (*ag*)

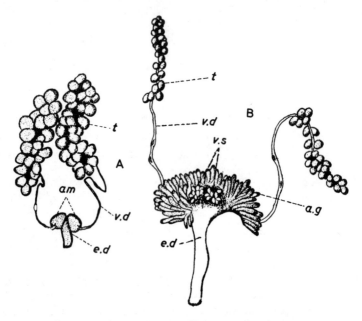

Fig. 46. *Blatta orientalis*, Male reproductive system

A, Reproductive system of 6th instar nymph. B, The same in adult male (both equally magnified). *am*, ampullae; *a.g*, accessory gland; *e.d*, ductus ejaculatorius; *t*, testis; *v.d*, vas deferens; *v.s*, vesicula seminalis. (*Adapted from* Qadri, 1938)

arise from the vesicle just mentioned and, since they are of mesodermal origin, they are classed as *mesadenia* (Fig. 46 B). In other insects from one to three pairs of accessory glands may occur, and those arising from the ectoderm of the ductus ejaculatorius are termed *ectadenia*. In many cases the accessory glands produce substances which go to form a kind of capsule or *spermatophore* that encloses the spermatozoa. It is deposited in the bursa copulatrix or vagina of the female during mating, the spermatozoa ultimately becoming free. Spermatophores vary in form and structure and several may be produced during a single mating. They occur in Orthoptera, Dictyoptera, Lepidoptera and other orders. In a few insects peptide secretions of the male accessory glands affect the female, reducing her readiness to mate again or stimulating her to lay eggs. *Vesiculae seminales* are found in many insects and are developments of the vasa deferentia. In *Blatta* they take the form of numerous outgrowths of the mesodermal vesicle (Fig. 46), and as the testes degenerate in adult cockroaches spermatozoa are then only found in the seminal vesicles.

Reproduction

In most insects reproduction depends on copulation between adults of opposite sex and the female then lays eggs, from each of which a single immature insect hatches after a more or less prolonged incubation period. Exceptions to these various generalizations, however, are not uncommon and are dealt with below.

Sperm Transfer. This normally occurs during copulation by methods indicated above, but a few insects show anomalous forms of sperm transfer. In many Apterygotes the males deposit semen externally in small droplets, which are then taken up into the genital tract of the female when she moves over them. The dragonflies are unique in that before mating the male transfers spermatozoa to a secondary copulatory organ near the front of the abdomen, from which the female receives them when the pair flies in tandem. In bed-bugs (*Cimex*) and a few related Heteroptera, the spermatozoa are transferred by the aedeagus, but pass through the abdominal body wall of the female into the haemocoele, thence migrating to the ovaries, where fertilization takes place.

Parthenogenesis. In this type of reproduction the eggs undergo full development without having been fertilized. It occurs in representatives of most orders and may be *obligate* – when males are absent or very rare and non-functional – or *facultative*, when it co-exists with normal bisexual reproduction. Four important types of parthenogenesis are known: (i) In the honey bee and some other insects the females lay two kinds of eggs. Those which are unfertilized have only the reduced (haploid) number of chromosomes and give rise exclusively to males, whereas the fertilized eggs, with the full diploid number of chromosomes, produce only females (including the sterile females or workers of the honey bee). The usual method of sex determination by sex chromosomes is therefore replaced by this more flexible method. (ii) In some saw-flies, some stick-insects and one species of scale-insect, the female again lays two kinds of eggs. Those which have been fertilized produce about equal numbers of males and females but in the un-fertilized eggs there is a fusion of the egg nucleus with the second polar body. This restores the diploid chromosome number but such eggs develop only into females. (iii) A striking type of obli-gate parthenogenesis occurs in some aphids, stick-insects, weevils and moths. The eggs are formed without meiosis and only female progeny result. Males do not occur in these species, which are able to reproduce rapidly but lack the genetic variability found in bisexual forms. (iv) In some aphids and Cynipid gall-wasps there is cyclical parthenogenesis: one or more parthenogenetic genera-tions alternates with a bisexual generation. In these cases par-thenogenesis occurs in the summer so that rapid reproduction can occur under favourable conditions while the advantageous genetic effects of bisexual reproduction are not entirely lost.

Viviparity. In some insects, embryonic development is com-pleted within the body of the female parent, which therefore produces living young instead of laying eggs. Such viviparity may mean little more than the retention within the vagina of otherwise normal and fully-yolked eggs until the young insects hatch out and are expelled. In tsetse flies (*Glossina*) and Pupiparan Diptera such as the sheep ked *Melophagus*, however, the larvae remain after hatching in the enlarged vagina of the female, where they feed and grow and are deposited as mature larvae ready to pupate (Fig. 47 D). In other cases (e.g., aphids and the ectoparasitic earwig *Hemimerus*) the eggs have no chorion and are practically devoid of yolk; a special placenta-like structure is therefore developed to nourish each embryo.

Paedogenesis. This term denotes reproduction by a juvenile stage. It occurs in only a few species, one of the best-known examples being the gall midge *Heteropeza* (Cecidomyidae). Here, as in the other known examples, reproduction involves both par-

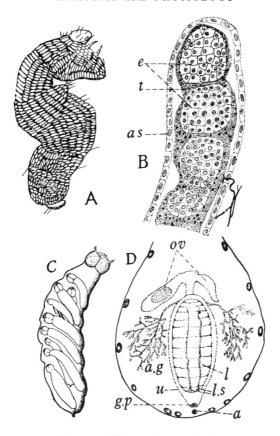

Fig. 47. Types of reproduction

A, remains of larva of *Autographa gamma* distended with pupae of the polyembryonic parasite *Litomastix* (*after* Silvestri). B, portion of a polyembryonic chain
derived from a single egg of a Chalcid parasite dissected from its host; *a.s*,
adventitious sheath produced by host; *e*, embryos surrounded by trophamnion, *t*
(*after* Marchal). C, paedogenetic larva of *Micromalthus* with daughter larvae ready
to emerge (*from a drawing by* H. E. Hinton). D, viviparity in *Melophagus*, ventral
aspect of abdomen seen as a transparent object, showing *l*, growing larva, in uterus,
u; *a.g*, accessory gland; *a*, anus; *g.p*, gonopore; *ov*, ovaries; *l.s*, larval spiracles

thenogenesis and viviparity. Within the body of the parent larvae
unfertilized eggs give rise to daughter larvae. These then eat their
way out of the body of the parent larvae which die during the
process. The daughter larvae may repeat this cycle or give rise to
normal male and female insects. Paedogenesis also occurs in the

complex life cycle of the beetle *Micromalthus* from N. America and S. Africa (Fig. 47 C).

Polyembryony. Polyembryony is the production of two or more – often very many – embryos from a single egg (which may be fertilized or develop parthenogenetically). It occurs in a Strepsipteran (*Halictoxenos*) and has been evolved in several more or less widely separated genera of parasitic Hymenoptera. The essential feature of the process is the separation of the blastomeres of the egg into groups of cells or morulae, each of which grows into an adult insect. Very early in development the egg develops a surrounding sheath or *trophamnion* (Fig. 47 B), which absorbs food material from the tissue fluids of the host and passes it on to the growing embryo. This covering accommodates itself to the increasing size of the polyembryonic mass which, in extreme cases, forms a tortuous chain of embryos filling the haemocoele of the host, which finally dies. The simplest cases of polyembryony are in *Platygaster* (a Proctotrupoid) whose species parasitize the Hessian fly and its allies. In *P. hiemalis* some of the eggs produce single embryos while the others divide and give rise to two individuals. In various other parasites embryonic fission results in the production of 8 or 10 to over 100 individuals from a single egg; they are always of the same sex. An extreme phase is reached in the Chalcid parasite, *Litomastix truncatellus*, of the common Silver Y moth, *Autographa gamma*, where the division of a single egg produces up to about 1000 individuals (Fig. 47 A).

Fecundity and the Equilibrium of Populations

Insects abundantly exemplify the rule that every living organism if allowed to multiply without restriction would sooner or later reach unprecedented numbers. Thus, the progeny of a single aphid – if all the members survived – would at the end of 300 days be somewhere in the order of the 15th power of 210. In nature, however, reproductive rate is rarely a criterion of the relative abundance of species. The Fulmar petrel does not breed until two or three years old and then lays only a single egg each year, yet it is believed to be one of the most abundant birds in the world. A parallel example is afforded by the tsetse flies (*Glossina*) which are often so abundant and harmful in tropical Africa. Each fly, however, produces but a single larva at a time and only a few in its whole life. In general, fecundity is adjusted to the probability of survival and is high where, as in most insects, the parent has no means for protecting the eggs or young. Thus among Tachinid flies (p. 195) the smallest number of eggs is laid by those species which deposit them, or the resulting larvae, on or within their hosts; in such species fewer than 200 eggs are produced. At the other extreme are those Tachinidae which lay 2000 to 6000 eggs upon or near the food of their hosts. Such eggs hatch only if they happen to be swallowed by the host. To revert to the tsetse flies, in these insects only a single egg matures at a time in the ovary and the resulting larva is nourished in the body of the female until ready for pupation. Evidently, many of the normal perils which beset the egg and larval stages are avoided and the swiftly flying inconspicuous tsete is well endowed for taking care of itself and its contained progeny.

The majority of species live in a state of *biological equilibrium* or *control*. The word control is ambiguous because in an economic sense control implies a level, often artificial, below that at which human interests, such as crops, are interfered with and this level may vary from year to year according to the state of the market or to variations in taste or fashion. The word equilibrium can also lead to misconception: the equilibrium is only relative to the very wide possible amplitude of natural fluctuations and may still comprise very considerable variations. But the meaning of equilibrium can be seen from two examples. The majority of British butterflies and moths still have about the same degree of commonness as was recorded by the collectors in the early nineteenth century; a few species only have been eliminated, chiefly by the wholesale destruction of vegetation, and a few have become commoner. Again, only a few kinds of British caterpillars, like the Green Oak Tortrix, ever become so common that the trees are defoliated and then only at rather infrequent intervals.

The factors which normally inhibit the great natural powers of increase are usually classified into *density dependent* and *density independent*. The latter are agencies which destroy on the average a percentage of a species which is irrespective of its numbers or reduce its reproductive rate by a fixed fraction. Climate in many of its effects is characteristically density independent. There is no agreement as to the most satisfactory detailed definition of density dependent but we can say in a general way that such factors adjust themselves to the numbers of the species so that they become more effective as it gets commoner. Examples of density dependent factors are competition within the species (for food, nesting sites, etc.; this is the most perfect example), competition with other species, and often the effects of parasites, predators and of diseases. It is probable that there are many less obvious examples, such as the tendency of a species when common to spread more extensively into unsuitable habitats and to leave, therefore, fewer progeny. In nature, both density dependent and independent agencies are always at work simultaneously but the former are essential for equilibrium since they alone adjust themselves to changing numbers. Moreover, they are often factors of a type more easily influenced by man than climate. One of the most familiar examples of density dependent control is the application by man of insecticides on occasions or in areas where harmful insects are abundant. A more spectacular example is the method known as *biological control*. This consists, typically, of the intentional introduction of an insect to control another insect or a weed. Well-known examples of the success of this method are the use of ladybirds (Coccinellidae) to control scale insects in California, and the destruction of prickly pear (*Opuntia*) in Australia by the larvae of the moth, *Cactoblastis*. In nearly all such examples, the harmful species was introduced into a new continent without its natural enemies. Later, these were searched for in the native land of the pest and introduced experimentally, often with striking results.

3
Development and Metamorphosis

EMBRYONIC DEVELOPMENT

The Egg. A typical insect egg (Fig. 48 A) is covered by an outer shell or *chorion* which is variously sculptured or ornamented. The chorion is secreted by the follicular epithelium and may consist of several layers; these include a very thin wax layer, which reduces water loss, and others containing protein but no chitin. Beneath the chorion is the delicate vitelline membrane, which is a product of the egg itself; further inner layers may be deposited after fertilization and during embryonic development. Insect eggs usually contain a large amount of yolk. This is a mixture of protein, lipid, carbohydrate and other materials lying within the meshes of a cytoplasmic reticulum. As in other Arthropods, the egg of insects is *centrolecithal*, a thin peripheral layer of cytoplasm – the *periplasm* – surrounding the yolk. One or more specialized pores or canals in the chorion are known as *micropyles*; through these the spermatozoa enter the egg at fertilization. In some insects the micropyles are also the channels through which oxygen diffuses into the egg; in other cases special respiratory channels occur separately and in some insects part or all of the chorion forms a plastron (p. 87), allowing respiration under water. *Maturation* of the egg usually follows the entry of spermatozoa. The egg nucleus moves towards the periphery and undergoes meiotic division, the polar bodies are segregated and the female pronucleus fuses with one of the spermatozoa, thus restoring the diploid chromosome number.

The Blastoderm and Germ-band. After fertilization, the

zygote nucleus divides repeatedly and many of the resulting cleavage nuclei pass outwards to form, with the periplasm, a continuous superficial cell-layer or *blastoderm*. Other cleavage nuclei, surrounded by adjacent cytoplasm, remain behind as *yolk cells*. While the blastoderm is developing, some of the dividing nuclei may pass to the posterior pole of the egg to form the primordial germ cells (Fig. 48 B); in other species, however, the germ cells are not segregated until later. As development proceeds the blastoderm thickens along the mid-ventral line, so giving rise to the *germ-band* (Fig. 48 C) which is destined to produce all the tissues of the embryo. The rest of the egg consists chiefly of yolk enclosed by the thin *extra-embryonic blastoderm*.

Developmental Physiology. Early embryonic development is controlled by three important centres; these are not visibly differentiated but their activities can be demonstrated experimentally. An anterior *cleavage centre* initiates nuclear division and migration. Near the posterior pole of the egg is the *activation centre*; this apparently produces a chemical substance which diffuses forwards and determines the formation of the germ-band. Only after this has taken place can the activation centre be eliminated experimentally without ill effects. Figs. 49 B and C, for example, show the respective effects of ligaturing off the posterior pole of the egg before and after the germ-band has been determined. In the position corresponding with the thorax of the future embryo lies the *differentiation centre*, from which visible differentiation of the germ-band proceeds, and which only becomes active after the product of the activation centre has reached it. An egg ligatured at an early stage anterior to the position of the differentiation centre may therefore develop into a dwarf embryo behind the ligature (Fig. 49 D). Before the work of the differentiation centre is complete the egg is capable of considerable 'regulation', i.e., parts of the egg can, under experimental conditions, develop into structures other than those to which they would normally have given rise. Later, the developing egg becomes 'mosaic' and localized injuries induced by, say, ultraviolet radiation causes corresponding defects in the larva. The time at which the developing egg passes from the regulation to the mosaic condition differs in different species. Thus, in the dragonfly *Platycnemis* regulation is possible up to the late blastoderm stage and an egg ligatured

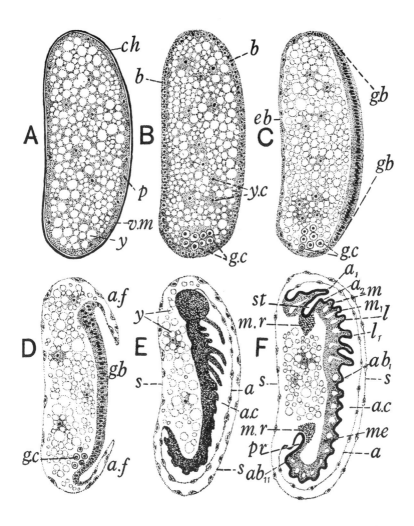

Fig. 48. Successive stages in the embryology of an insect

A, cleavage and migration of nuclei to the periplasm (*p*). B, formation of blastoderm (*b*). C, development of the germ-band (*g.b*). D, developing amniotic folds (*a.f*). E, embryo enclosed in the amniotic cavity (*a.c*). F, section through the same. Other lettering: *a*, amnion; a_1, 1st antenna; a_2, appendage of intercalary segment; ab_1, 1st abdominal appendage; ab_{11}, 11th do.; *ch*, chorion; *eb*, extraembryonic blastoderm; *g.c*, primordial germ cells; *l*, labium; l_1, 1st leg; *m*, mandible; m_1, maxilla; *me*, mesoderm; *m.r*, mesenteron rudiment; *pr*, proctodaeum; *s*, serosa; *st*, stomodaeum; *v.m*, vitelline membrane; *y*, yolk; *y.c*, yolk cells. (In B–F the chorion and vitelline membrane have been omitted)

during the period of regulation may produce two dwarf embryos (Fig. 49 E). In *Drosophila* and other Diptera, however, the egg has already reached the mosaic stage by the time it is laid, while in the honey bee this condition is achieved at a stage intermediate between those of *Platycnemis* and *Drosophila*. A consequence of the early but histologically unrecognizable differentiation of the embryo is that the blastoderm can be divided into a number of 'presumptive areas', each normally giving rise to a particular germ layer or region of the later embryo. Such blastoderm 'fate maps' show a general similarity within the Pterygotes and Thysanura, but an appreciably different pattern occurs in the more distantly related Collembola and Diplura.

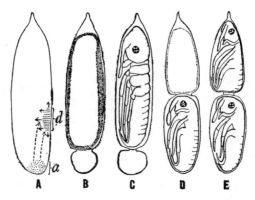

Fig. 49. Regulation of development in *Platycnemis*
For explanation see text. *a*, *d*, positions of activation and differentiation centres,
(*Based upon* Seidel)

Embryonic Membranes and Gastrulation. Sooner or later the germ-band becomes enclosed by *amniotic folds* that arise from its edges. These folds grow towards one another so as to meet and fuse, thus enclosing the germ-band within a space known as the *amniotic cavity* (Fig. 48 D, E; Fig. 50). Of the covering membranes thus formed the outer layer or *serosa* is continuous with the *extra-embryonic blastoderm* and the inner membrane or *amnion* is continuous with the margins of the germ-band. These two membranes, and the cavity they enclose, function as an insulating cushion which protects the growing embryo from injury.

During growth of the amniotic folds *gastrulation* takes place as

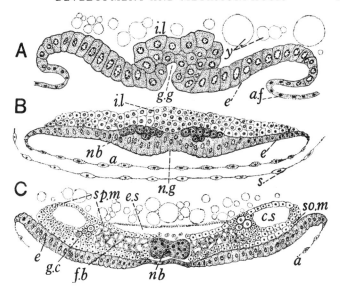

Fig. 50. Transverse sections through the germ-band of an insect at successive stages of development

A, formation of gastral groove (*g.g*) and inner layer (*i.l*). B, beginnings of neural groove (*n.g*) and ventral nerve cord (*n.b*). C, development of coelom sacs (*c.s*) and beginning of separation of the embryo from the yolk resulting in the formation of the epineural sinus (*e.s*). Other lettering: *a*, amnion; *a.f*, amniotic fold; *e*, ectoderm; *f.b*, fat-body; *g.c*, primordial germ cells; *s*, serosa; *so.m*, somatic layer of mesoderm; *sp.m*, splanchnic layer of mesoderm; *y*, yolk

a ventral furrow-like ingrowth on the middle line of the germ-band. It begins at the site of the future stomadaeum and gradually extends to the caudal end of the germ-band, whose cells thus become deployed as a lower or *inner layer* beneath the outer layer or *ectoderm* (Fig. 50). In most insects the inner layer gives rise to mesodermal structures and rudiments of the mid gut, but the latter may arise in various ways and this has led to a less strict interpretation of embryonic germ layers than was formerly the case.

Segmentation. Very early in development the two-layered germ-band or embryo, as it may now be called, becomes divided by transverse furrows into a series of segments which ultimately number 20 in all. Segmentation is a gradual process beginning anteriorly and extending backward (Fig. 51). The embryo is at first divisible into a *protocephalic* or *primary head region* and a *proto-cormic* or *primary trunk region*. As development progresses the first three protocormic segments become added to the protocephalic region. The next three body segments are grouped to form the thorax and the remaining eleven segments, together with the non-segmental telson, constitute the abdomen. Each segment except the first typically develops a pair of outgrowths or *embryonic appendages* (Fig. 51). The first or pre-antennal segment is formed of the large *procephalic lobes*. The first pair of appendages or antennae belongs to the second segment, while the very small

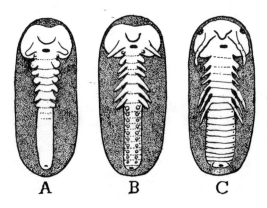

Fig. 51. Ventral side of developing embryos in A, *Protopod*, B, *Polypod*, and C, *Oligopod* stages

second pair is transitory and soon disappears. The third, fourth and fifth pairs grow respectively into the mandibles, maxillae and labium. The next three pairs of embryonic appendages are usually larger and more conspicuous; they are the forerunners of the thoracic legs. Finally, there follow eleven pairs of abdominal appendages of which the last pair becomes the cerci and the remaining pairs are usually resorbed before hatching. The presence of these evanescent limb rudiments can best be interpreted as indicating a many-legged ancestral stage. The number of ab-dominal segments varies in the adults among different groups of insects since some of the anterior and posterior segments tend to disappear.

Mesoderm. The cells of the inner layer become arranged, for the most part, in two longitudinal bands which are shortly afterwards marked off into divisions or segments corresponding with those bearing the appendages (Fig. 48). Most of these mesodermal segments acquire, in generalized insects, a pair of cavities or *coelom sacs*. The dorsal or splanchnic walls of these sacs give rise to the gonads, visceral muscles and fat-body, while the ventrolateral or somatic walls produce the muscles of the body and the appendages. The *body cavity* begins as the *epineural sinus* which is mainly formed by the separation of the yolk from the embryo in the mid-ventral region (Fig. 50 C). This process extends laterally and upward, and results in the epineural sinus and most of the coelom sacs becoming confluent, thus forming the permanent body cavity or haemocoele. The *primordial germ cells* soon migrate forward and, after separating into two groups, they penetrate into the gonad rudiments, where they become established.

Nervous System. Shortly after gastrulation the central nervous system develops as a pair of longitudinal *neural ridges* of the ectoderm separated by a median *neural groove* (Fig. 50 B). The neural ridges become segmentally constricted into *neuromeres* or primitive nerve ganglia, while their intersegmental portions give rise to the connectives. It will be noted that the whole of the nervous system and the sense organs are ectodermal in origin and that their rudiments become separated from the outer ectoderm which forms the body wall (Fig. 50 C). The ganglia of the acron and of the first three head segments amalgamate to form the *brain*, while the succeeding three cephalic neuromeres fuse to become the *suboesophageal ganglion*. The neuromeres that follow develop into the thoracic and abdominal ganglia.

Alimentary Canal. An ingrowth of the ectoderm, just behind the antennae, forms the *stomodaeum* or embryonic fore intestine and a corresponding posterior ingrowth or *proctodaeum* gives rise to the hind intestine (Fig. 48 F). The *mesenteron rudiments* usually arise as groups of cells closely associated with the stomodaeal and protodaeal ingrowths; these cells multiply, grow towards each other and finally enclose the yolk in the form of a complete tube – the *mesenteron* or embryonic mid intestine. By the disappearance of the walls separating the mesenteron from the stomodaeum and proctodaeum respectively, a through passage is established in the alimentary canal. The *Malpighian tubules* arise as outgrowths of the proctodaeum, close to its union with the mesenteron. Though often regarded as ectodermal in origin, there is some reason to think that they are endodermal or arise from an undifferentiated meristematic zone. The *salivary glands* arise as paired ectodermal ingrowths at the sides of the labial segment.

Tracheal System. The tracheae develop from paired lateral ingrowths, near the bases of the appendages, on the meso- and metathorax and on the first eight abdominal segments. The mouths of these invaginations become the *spiracles* and, at their inner ends, anterior and posterior longitudinal extensions meet and fuse to form the main tracheal trunks.

Later Phases of Development. The embryo always forms on the ventral surface of the egg, but in the lower insects it becomes invaginated into the abundant yolk that is present in these forms. This process occurs through the embryo traversing an arc so that its ventral surface now faces the dorsal side of the egg. After a short time it begins to reverse its position and ultimately regains the ventral side of the egg. These movements of the embryo during development are termed *blastokinesis*, but the significance of the process is obscure. The germ-band forms the ventral part of the developing insect and, in order to complete the embryonic body, its margins begin to grow upward. The final result is the completion of the embryo on the dorsal side though the details of the process differ among different insects. The upward growth involves not only the ectoderm or body wall but also the epineural sinus and the mesoderm, while the developing mid intestine ultimately encloses the yolk. The embryonic membranes later rupture and, becoming contracted, are finally resorbed.

POSTEMBRYONIC DEVELOPMENT

Hatching. When ready to emerge from the egg an insect has to force its way through the chorion in order to reach the outer world. In many cases the chorion is torn open by means of provisional structures known as hatching spines or egg bursters. These are located on the head, or other parts of the body, where they may

remain until the insect has undergone its first moult. In other
instances, notably among Mallophaga and Hemiptera-Heteroptera,
a preformed egg cap or operculum is pushed open to allow the
insect to emerge. The force used in hatching is chiefly muscular
activity but, as a preliminary, an insect may swallow air or the
amniotic fluid and the resulting increase in bulk and turgidity play
their part in the process.

Growth. Insects grow in cycles which alternate with the periods
when moulting takes place (see below). The tissues may grow
through cell multiplication, through an increase in the size of the
individual cells or through both processes occurring together. In
Endopterygote insects it is often found that those tissues which
grow mainly by cell multiplication pass over from larva to adult
with little change, while those which develop by cell hypertrophy
break down in the pupal stage and are replaced by new adult
tissues. The soft cuticle of some insects stretches considerably
during growth but more strongly sclerotized parts such as the head
capsule grow discontinuously, the enlargement becoming apparent
after each moult. Such structures often increase in size by an
approximately constant ratio at every moult (Dyar's law). If, however, there is an unusually long or short interval between two successive moults then the ratio may be proportionately larger or
smaller than normal. The various parts of the body tend to grow
at rates which differ from each other and from the growth rate
of the whole body. This often results in allometric growth; the
size of a part is then a constant power function of the size of the
whole body, so that when the two dimensions are plotted against
each other on logarithmic scales a straight line is obtained.

Moulting. In order to accommodate the growing tissues, immature insects shed their cuticle at intervals; this process is known
as moulting and the cast skin forms the *exuviae*. Before it moults,
the insect stops feeding and may become quiescent for a short
time. Meanwhile the epidermal cells enlarge and may divide
mitotically, the old cuticle becomes detached from the epidermis
(a process known as *apolysis*) and the latter begins to lay down
a new cuticle. The space between old and new cuticles then
becomes filled with the moulting fluid which is secreted by the
epidermis and contains two important enzymes, a protease and a

chitinase (Fig. 52). These dissolve the old endocuticle and the products of digestion are absorbed through the epidermis; the newly formed cuticle remains unaffected by the enzymes. What is left of the old cuticle is now ready to be shed, an act that may be referred to as *ecdysis*. The insect contracts its abdominal muscles and increases the pressure of blood in the head and thorax. It may also distend itself by swallowing air, or water in aquatic forms, and the resulting forces rupture the old cuticle along predetermined lines of weakness – the ecdysial cleavage line of the head and a median dorsal line in the thorax. When the old

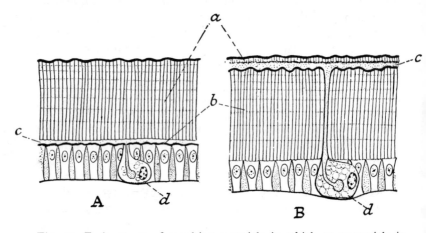

Fig. 52. Early stages of moulting a cuticle in which an exocuticle is wanting

A, new epicuticle formed; digestion of old endocuticle scarcely begun. B, digestion and absorption of old endocuticle almost complete. *a*, old cuticle; *b*, new cuticle; *c*, moulting fluid; *d*, dermal gland. (*After* Wigglesworth)

skin has split the insect gradually struggles out, often aided by gravity, since many insects suspend themselves head downwards when moulting. In issuing from the old skin, the insect withdraws its limbs from their former coverings and the old linings of the tracheae, fore gut and hind gut are left behind with the exuviae. The newly moulted (*teneral*) insect has a soft, flexible, lightly pigmented cuticle; it swallows air (or water) and so again increases its volume. Many of the muscles remain contracted for a time and in this way blood pressure expands the wings and other appendages to their full size. Finally, the cuticle hardens and darkens though it may continue to increase in thickness for a considerable time.

The intervals between one moult and the next are sometimes known as stages or *stadia* and the form assumed by the insect in any particular stadium is called an *instar*. Thus, the insect hatches from the egg as the first instar; at the end of the first stadium it moults into the second instar, and so on. For some time before ecdysis the new instar lies within the old but unshed cuticle and is then known as a *pharate instar*. The number of instars differs greatly in different groups of insects, but is often approximately constant within a group. In some relatively primitive insects like the Ephemeroptera and Plecoptera there may be more than 20 preimaginal instars; in the Cyclorrhaphan Diptera there are four, and in the Lepidoptera from four to ten. In many Apterygota (Thysanura, Collembola, *Campodea*) the insects continue to moult after reaching sexual maturity but in the Pterygota the adults never do so. The hormonal control of moulting is discussed on p. 129.

Metamorphosis. The appearance and structure of the newly hatched insect differs more or less extensively from that of the adult. Its development therefore involves changes of form or *metamorphosis*, which may include the loss of purely juvenile structures as well as the differentiation of distinctively adult organs. The magnitude of these changes varies from group to group, but two main types of metamorphosis may be distinguished: incomplete metamorphosis – also known as direct or *hemimetabolous* development – and complete metamorphosis (indirect or *holometabolous* development).

Hemimetabolous development occurs in the Apterygota and almost all Exopterygota (Fig. 53). The immature stages, sometimes referred to as *nymphs*, usually resemble the adults in habits and in many structural features, and metamorphosis consists mainly of a relatively gradual differentiation of adult characters. The wings and external genitalia usually make their appearance at an early stage as external rudiments which increase in size and complexity with each successive instar. The simple gonad rudiments of the young nymph gradually differentiate and grow and the efferent ducts of the reproductive system develop progressively. The number of segments in the antennae, tarsi and cerci may increase with successive moults and the compound eyes grow

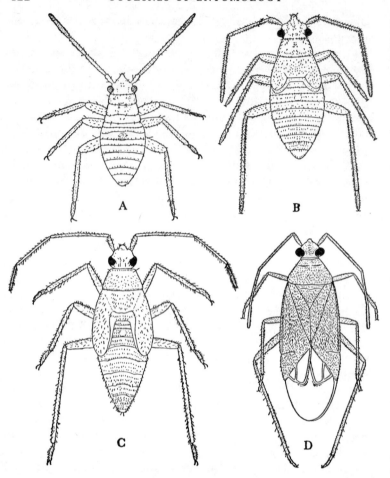

Fig. 53. Incomplete metamorphosis of an Hemipteran (*Lygocoris pabulinus*)

A, nymph, 2nd instar. B, do., 4th instar. C, do., 5th instar. D, imago. (*From* Petherbridge and Thorpe)

with the differentiation of additional ommatidia. There are often changes in coloration, in the proportions of various structures, and in the number and distribution of cuticular bristles. The changes occurring at the moult into the adult are sometimes rather more striking than those at earlier ecdyses. This is especially obvious in the Ephemeroptera, Odonata and Plecoptera, where the young stages are aquatic, with gills and other structures which are lost on transformation into the terrestrial adults.

Holometabolous development occurs in virtually all the Endopterygota. Here there is a series of active, feeding larval instars – which usually resemble each other closely, but may differ enormously from the adult in habits and structure – followed by a single quiescent pupal instar, which does not feed and which finally moults into the adult. Compared with the hemimetabolous insects, holometabolous ones show a marked suppression of adult features in the larval stages, combined with a strong tendency towards the evolution of specifically larval organs which are lost at the change into the pupa. The complex internal changes which accompany pupation are described on p. 127.

A few Exopterygote insects – the Thysanoptera, Aleyrodidae and male Coccoidea – have a metamorphosis which resembles that of the Endopterygota. The earlier instars differ appreciably from the adult and the life-cycle includes a period of extensive structural change comparable to that found in the Endopterygote pupa – the Thysanoptera and male Coccidea have, in fact, two or, in some Thysanoptera, three pupal instars.

The Endopterygote Larva. The larvae of holometabolous insects display an enormous range of structural variation and are adapted for life in a very wide variety of environments. In general, however, they may be distinguished from the immature stages of Exopterygote insects by a number of external features. There are no compound eyes, but their place is often taken by one or more pairs of lateral ocelli; dorsal ocelli are absent. External wing pads and genital appendages are absent, though internal rudiments usually occur beneath the cuticle of at least the older larvae. Finally the general integument is commonly less sclerotized and in some cases the head capsule, cephalic appendages and legs are greatly reduced or absent.

Endopterygote larvae may conveniently be divided into a number of different types (Fig. 54) though intermediate forms are not uncommon. The highly specialized *protopod* type (Fig. 54 A) occurs in the early instars of a few endoparasitic Hymenoptera and Diptera. It resembles the early embryonic stage of many insects and is little more than a precociously hatched embryo, with its ill-defined segmentation, its limbs rudimentary or absent and its

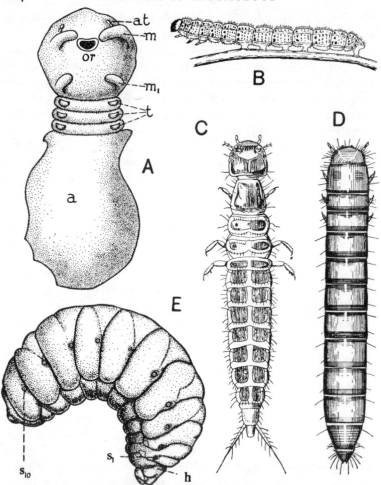

Fig. 54. Types of Endopterygote larvae

A, protopod (Proctotrupoidea). B, polypod (Lepidoptera). C, oligopod (Coleoptera, Staphylinidae). D, do. (*Tenebrio*). E, apodous (Hymenoptera: honey bee). (A *from* Kulagin, E *from* Nelson.) *a*, abdomen; *at*, antenna: *m*, mandible; *m₁*, maxilla: *h*, head; s_1, s_{10}, spiracles; *t*, thoracic limbs; *or*, mouth

incompletely differentiated internal organs. The *polypod* type, typified by the Lepidoptera (Fig. 54 B) and many sawflies, is a structurally more advanced form provided with a number of abdominal processes which may be used in locomotion, though their serial homology with the thoracic legs is doubtful. The *oligopod* type (Fig. 54 C, D) is the prevailing form in the Coleoptera and Neuroptera. It has a well-developed head, antennae and thoracic legs but no abdominal processes except sometimes for a pair of cercus-like appendages. According to their general facies, the members of this group may be subdivided into several forms, of which two are worth special mention. *Campodeiform* larvae (Fig. 54 C) are usually active predators with relatively long legs and often a somewhat depressed, sclerotized body with a prognathous head and powerful mouthparts. The more reduced *scarabaeiform* larvae, found in the cockchafers and some other beetles, are characteristically C-shaped, with shorter legs, less highly developed mouthparts and a more membranous integument (Fig. 80 D). The last major type is the *apodous* larva which lacks all legs or similar processes (Fig. 54 E). Such larvae usually live among abundant food and are characteristic of the Diptera, the aculeate and parasitic Hymenoptera and, among Coleoptera, of the weevils. According to the extent to which the head capsule is reduced and retracted into the thorax, apodous larvae may be further subdivided into *eucephalous*, *hemicephalous* and *acephalous* types (p. 191).

Towards the end of the last larval instar, the insect becomes less active, stops feeding and prepares for the moult into the pupa. It may construct a cocoon or pupal cell (see below) and as the pupal organs develop within the larval integument its body often becomes somewhat contracted and depressed. This phase of development is sometimes known as the prepupa but it is more accurately described as the pharate pupa (cf. p. 121).

The Endopterygote Pupa. This instar differs from the larval phase in many respects. It is more or less quiescent, it does not feed and it is the site of more or less profound internal changes which transform the larval organization into that of the adult (pp. 127–8). Externally, the pupa is usually rather like the adult in form; it has well-developed external wing pads, thoracic legs and antennae but the cuticle is often thin and lightly pigmented. It is a vulnerable stage in the life cycle, with few special means of defence, and is therefore often concealed in the soil, in debris or in crevices. It may be enclosed in a silken cocoon, spun by the larvae from the labial glands (many Lepidoptera) or Malpighian tubules (many Neuroptera). In other cases a pupal cell is constructed from particles of soil, chips of wood or other debris, held together by

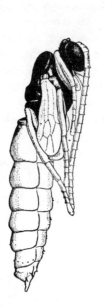
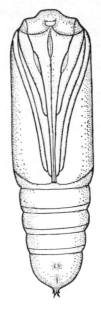

Fig. 55. Exarate or free pupa of an Ichneumonid (Hymenoptera)

Fig. 56. Obtect pupa of a Noctuid (Lepidoptera)

Fig. 57. Puparium of a Cyclorrhaphan Dipteran (Muscidae): Dorsal view

various secretions and in the Cyclorrhaphan Diptera the persistent cuticle of the last larval instar is transformed into a hard, dark, ovoidal *puparium* which protects the delicate adecticous exarate pupa inside it (Fig. 57). Freely exposed pupae, such as occur in the ladybirds (Coccinellidae) and some Lepidoptera, are often protectively coloured and have a rather thick integument. For some time before the pupal cuticle is shed, the fully formed adult lies within it; these older 'pupae' are sometimes able to move actively (e.g., in the Trichoptera), but strictly speaking it is the pharate adult which is responsible for the movements.

Two main types of pupae may be recognized. The *decticous* pupa, found in the Neuroptera, Mecoptera, Trichoptera and a few primitive Lepidoptera and Hymenoptera, is provided with functional mandibles which are used by the pharate adult to break out of the cocoon or pupal cell before the adult emerges. The *adecticous* type of pupa, found in the other Endopterygotes, has no functional mandibles. It may escape from the cell with the aid of various spines or other special processes or, as in the Coleoptera and Hymenoptera, it moults into the adult within the cell. Adecticous pupae may, like all decticous ones, be *exarate*, with the appendages projecting freely from the body (Fig. 55) or they may be *obtect*, as in most Lepidoptera, with wings, leg, antennae and mouthparts firmly soldered down to the body (Fig. 56).

Internal Metamorphosis. In hemimetabolous insects the development of external and internal structures is gradual and direct. The organs of the nymph generally become transformed into those of the imago with little change beyond increase in size, alterations of proportion and limited structural elaboration. In the holometabolous life cycle, however, the onset of pupation marks the beginning of a variable, though often extensive, destruction of larval organs and tissues (*histolysis*), accompanied by a rapid equivalent differentiation of adult structures (*histogenesis*). These processes continue throughout the pupal stage and may not be fully completed until after the adult has emerged.

In the more primitive Endopterygota such as the Neuroptera, and also in the Coleoptera, relatively little histolysis occurs and a considerable proportion of the larval tissues and organs pass over to the adult with only slight modifications. In the Hymenoptera and Diptera, on the other hand, the epidermis, alimentary canal, salivary glands and many muscles of the larva are replaced by entirely new adult formations, and in some cases the fat-body, Malpighian tubules and heart also undergo the same fate. Even those structures which are not histolysed may experience considerable histological redifferentiation.

Histolysis begins with the death of the tissues concerned, which may then break up and dissolve without the intervention of phagocytic blood-cells. Often, however, the latter engulf and digest the disintegrating tissue fragments or even attack the recently dead tissues by invading them while they retain much of their structural integrity. The breakdown products of histolysis accumulate in the blood where they provide materials used in histogenesis. The cells from which the adult tissues develop seem to have retained a latent capacity for imaginal differentiation which has been largely or entirely suppressed during the larval phase. In some cases these cells remain histologically indistinguishable from neighbouring cells for much or all of the larval phase; in others they become recognizable as *histoblasts* which are sometimes present from the embryo or from some later larval stage as organ rudiments or *imaginal disks* or *buds* (Fig. 58). The imaginal disks of the wings, legs, mouthparts, antennae and external genitalia arise as slight folds or thickenings of the epidermis which may come to lie in sac-like invaginations (Fig. 58 c). Their growth and differentiation accelerate rapidly as the larval period comes to an end and they are everted at pupation. The imaginal buds of the epidermis, gut,

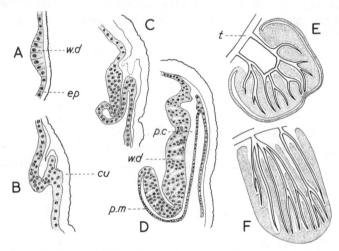

Fig. 58. Imaginal disks of wing in a moth (*Ephestia kuehniella*)
A, B, C, D, sections through successively older wing disks; E, F, surface view of younger and older wing disks showing tracheal supply. (After Köhler). *cu*, cuticle; *ep*, epidermis; *p.c*, peripodial cavity; *pm*, peripodial membrane; *t*, tracheal trunk at base of wing disk; *w.d*, wing disk.

glands, main tracheal trunks, etc., are nests of cells localized in the walls of the larval organs; they proliferate actively in the pupa, supplementing or replacing the larval tissues.

Hypermetamorphosis. Insects in which two or more of the successive larval instars differ widely in form are said to undergo hypermetamorphosis. The process is characteristic of certain parasitic groups and the alterations in larval form are accompanied by changes in their mode of life. Thus, in the Meloidae or oil beetles the first instar is an active hard-skinned campodeiform larva which is transported to the nest of its host, usually a bee, and there moults first into a soft-bodied, short-legged larva and later into one or more fleshy grub-like instars without legs or with vestigial ones. Hypermetamorphosis also occurs in all Strepsiptera, in the Neuropteran family Mantispidae and in many parasitic Hymenoptera (where the first-stage larvae are very variable in appearance).

Hormonal Control of Moulting and Metamorphosis. Growth, moulting and the realization of imaginal characters during postembryonic life are under the control of a balanced system of hormones produced by endocrine glands in the anterior part of the body (p. 97). The processes involved seem to be fundamentally similar in both the Exopterygote and Endopterygote insects. The neurosecretory cells of the brain produce a hormone which activates the thoracic glands or their homologues and causes the latter to secrete a second hormone or a mixture of similar hormones. It is these, which have now been synthesised chemically and are sometimes known as ecdysone and its derivatives, which induce the insect to moult. While the insect is still young, however, a third group of hormones, produced by the corpora allata and known as 'juvenile hormones' are also present in sufficient quantity to inhibit the development of imaginal features. The moults induced by the thoracic gland hormone therefore result in a sequence of larval or nymphal instars. Later in larval or nymphal life the production of the juvenile hormone ceases or is reduced in a more or less abrupt fashion and the previously inhibited adult structures can now develop. Quantitative differences in the changing balance between the hormones of the thoracic glands and the corpora allata thus seem to underlie the difference between the relatively sudden transition to the imago in holometabolous insects and the more gradual change in hemimetabolous ones. Experimental proof of the above explanation has been obtained in many ways. For example, young insects deprived of their corpora allata moult into miniature adult-like forms while the implantation of active corpora allata into late nymphs or larvae leads to the production of supernumerary immature instars.

Diapause. In some insects, under certain conditions, any one of the developmental stages – egg, nymph, larva or pupa – can pass into a more or less prolonged state of arrested development or *diapause*; even the adult may enter what has been called 'reproductive diapause', during which the reproductive organs remain non-functional. Development cannot then be resumed, even under apparently favourable conditions, until the diapause has been 'broken'. Various environmental changes are now known to be responsible for inducing diapause, though they normally exert their effect some appreciable time before development is arrested. The most important environmental influence seems to be the number of hours of daylight per day; a regime of short day lengths (8–12 hours of light per day) generally induces diapause but in some insects an opposite relationship holds good. In temperate climates diapause commonly occurs in the overwintering developmental stage and is usually broken through exposure to a period of low temperature. The immediate cause of diapause in the nymph, larva or pupa is probably a temporary lack of the thoracic gland hormone necessary for moulting and growth. In the silkmoth *Bombyx*, however, the eggs may be induced to enter diapause by a hormone which had previously been released into the blood of the female parent from neurosecretory cells in the suboesophageal ganglion. From a biological standpoint, diapause is an adaptation which often permits survival without feeding during adverse environmental conditions and which tends to synchronize development so that all members of an insect population resume activity together when conditions become favourable.

4

Some Important Modes of Life in Insects

Aquatic Insects. Some of the best examples of adaptation to a particular mode of life are afforded by aquatic insects. Such insects have, in all cases, secondarily acquired this mode of life and their degree of adaptation varies within wide limits. The vast majority of aquatic insects inhabit fresh water; a much smaller number occur in brackish waters, while very few have colonized the sea. A midge, *Pontomyia*, in Samoa spends its whole life, even in the adult stage, under the sea.

Adaptations to an aquatic life have arisen independently in the most diverse orders of insects. Thus, among beetles and water-bugs many species are aquatic both in their immature and adult stages; the Trichoptera and many Diptera are aquatic only as larvae and pupae, while the Plecoptera, Odonata and Ephemeroptera, with one exception in each of the first two orders, live in the water during their nymphal instars. The main adaptive features affect the form, relations with the surface film, methods of feeding and locomotion and, most important of all, respiration.

The highly polished, smooth, elliptical contour of many water beetles serves to reduce resistance to the water during swimming. The nymphs of mayflies, and the larvae of certain Diptera and Coleoptera that inhabit torrential streams assume a greatly flattened form. This feature, coupled with the provision of special anchoring devices (spines, suckers), helps to prevent such creatures from being swept away by the current.

Many aquatic insects have to pierce the surface film in order to maintain their posterior spiracles in contact with the atmosphere and, at the same time, prevent water from entering the trachea.

In some cases (e.g., mosquito larvae) perispiracular glands provide an oily secretion that imparts hydrophobe properties to the surrounding cuticle, i.e., the water retreats from such areas, leaving the surface dry. In other cases special devices of hairs achieve the same object. Thus, when the larva of *Dytiscus* is supported at the tail extremity by the surface film the entry of water into the tracheal trunks is precluded owing to the presence of a circlet of hydrofuge hairs around the posterior spiracles. These hairs are so closely set that water is unable to penetrate between them when they pierce the surface film. They function in such a manner that their outer surfaces show strong hydrophile properties, while the inner surfaces of these hairs are hydrophobe in character. In the Collembolan *Podura aquatica* and in the water skaters (*Gerris*) a coating of fine hydrophobe hairs covers the body and holds the water at a distance so that these insects are incapable of being wetted.

Special adaptations concerned with feeding are evident among predators. These are discussed on pp. 133–4 and include prehensile fore legs, sharply toothed projecting jaws as in the *Dytiscus* larva and the labial 'mask' of the Odonata. Larvae of *Simulium* and of mosquitoes are provided with vibratile mouth brushes that set up water currents and thereby waft their microscopic food into the gullet.

In many aquatic insects the third pair of legs, or the second and third pairs, are specially modified for swimming. Thus, in *Dytiscus* the hind legs are elongated and much flattened from side to side so as to function as oars. The tibia and tarsus bear closely set fringes of swimming hairs along their upper and lower edges. These hairs become spread during the swimming stroke and fall back when the leg is drawn forward. The tarsus rotates on its axis in a way which allows the swimming stroke to be made by its broad surface, while its edge cuts through the water in reaching the return position; in other words the rotation is comparable with a sculler 'feathering' his oars. *Notonecta* (Fig. 75 F) and its allies, like *Dytiscus*, swim by means of their oar-like hind legs which operate simultaneously. *Hydrophilus*, on the other hand, uses its hind legs alternately while swimming in much the same sequence as a walking insect. The Dytiscidae consequently swim a long straight course, whereas *Hydrophilus* pursues a somewhat wobbling and less efficient mode of progression. Whirligig beetles (*Gyrinus*) perform their rapid gyrations on the surface of ponds or slowly-moving streams and swim by means of their greatly modified second and third pairs of legs. Some fly larvae, notably those of *Chironomus* and mosquitoes, swim by vigorous muscular action of the abdomen that results in side-to-side wriggling movements through the water. Certain minute Chalcids (*Caraphractus*, etc.), that parasitize the eggs of various aquatic insects, swim beneath the water by means of their wings.

The transparent or phantom larva of *Chaoborus*, which is a close ally of mosquitoes, has its tracheal system mainly represented by two pairs of bean-shaped sacs that act as hydrostatic organs (Fig. 59). This larva is able, by a chemical process which contracts or expands their walls, to vary the size of the sacs so that its buoyancy can be adjusted to the pressure of the water at different depths.

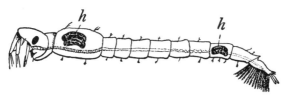

Fig. 59. Larva of *Chaoborus*, showing hydrostatic organs, *h*

The respiratory adaptations of aquatic insects are discussed in the chapter dealing with the general subject of respiration (p. 86).

Feeding Habits of Insects. Insects obtain their energy from food in which it has directly or indirectly been stored by green plants. These are the necessary intermediaries between other living organisms and the sun. It is not surprising, therefore, that a majority of insects feed on green plants and another large contingent on dead or decaying plant substances. It is the plant-feeding insects such as locusts, greenfly and caterpillars which are the most familiar enemies of the farmer, while other insects which eat wood, grain or tobacco are pests of urban civilization. A substantial number of insects, however, feed on substances which have been further elaborated by other animals from the primary sources. Such are the carrion beetles, some of which also damage furs; the clothes moths which, in the wild state, eat wool and feathers in birds' nests, and the numerous species which develop in the dung of vertebrates. Two types of specialized insect behaviour which are of great interest and of considerable importance to man are predatism – the killing and eating of other animals, especially insects – and parasitism – where the victim is either not killed at all or at least survives for a considerable period.

Predatism. A predator is an animal of carnivorous habit that seeks out and devours its prey. This kind of behaviour prevails among very diverse groups of insects and the prey consists of members of their own or other classes of invertebrates. In conformity with this mode of life certain structural adaptations are manifested in (1) the legs, (2) the mouthparts and (3) the sensory organs. In some cases adaptive changes in behaviour are also evident.

Many predators use their legs for seizing their victims and for holding them while they are being devoured. Thus, in dragonflies and adult Diptera of the family Asilidae, or robber-flies, all three pairs of legs are used for this purpose, these limbs being notably elongate and spiny. In the water-bugs (*Notonecta*, etc.) the fore legs alone are adapted for seizing and holding the prey: they are held well in front of the head and terminate in sharp prehensile claws. In the Mantidae, and in the Neuropterous family Mantispidae, the prey is caught and impaled by means of the combined femora and tibiae of the fore legs. It will be noted that the femur bears a ventral channel flanked by a double series of spines. The tibia is adapted to close in this groove after the manner of the blade of a pocket-knife, its sharply toothed edge acting in conjunction with the femoral spines in impaling the prey. In these two families, and in various others that use the fore legs for capturing their prey, the coxae of the limbs concerned are notably elongated (Fig. 5 D) so as to throw the legs forward and also allow of increased freedom of movement. Other predators have projecting sharply pointed mandibles for seizing their victims, as is well shown in larvae of ground beetles (Fig. 80), of *Dytiscus* and of most Neuroptera (Fig. 78). In dragonfly nymphs, as already noted, the labium is modified into a prehensile organ or mask (Fig. 67 D).

The sensory organs of many predators are highly developed, especially the compound eyes – a feature well seen, for example, in the Mantidae, Odonata, some Carabidae and Asilidae. When the prey are few and far between, and therefore require agile search, the legs of predators are adapted for running (Carabidae); or the predators are strong rapid fliers (Odonata, Asilidae). In cases where the specific prey is abundant and sedentary the legs and sense organs of such predators are not specially developed, since in most cases very little searching for their victims is required. Larvae of Neuroptera and of hover-flies or Syrphidae, together with the Coccinellidae, come under this category and they develop from eggs laid in close proximity to the aphids, etc., that form their food.

The mandibles, and sometimes the maxillae also, of predators are sharply pointed and adapted for piercing; they may also be toothed as in Odonata, being thus adapted for lacerating and tearing. In the Asilidae, together with other predacious flies, and also in many waterbugs, the mouthparts are developed into a rigid horny proboscis.

A considerable number of predators lie in wait for their prey, suddenly pouncing upon it when within reach. This adaptive habit is well seen among dragonfly larvae and Mantidae; the members of these two groups are usually cryptically coloured and thereby allow unwary prey to come within easy reach. Others, such as the larvae of Tiger Beetles (Carabidae, Cicindelinae), and those known as ant-lions (Myrmeleontidae), construct pits to ensnare their prey, while carnivorous Trichopterous larvae, that bear no cases, make web-like snares for entangling their victims.

Parasitism. A parasite is an organism that lives temporarily or permanently in intimate relationship with some other organism (its host) and at whose expense it obtains its nourishment. Among insects parasites fall into two very different categories, viz. true, or 'non-fatal', parasites, and 'fatal' parasites or 'parasitoids'.

Mallophaga and Siphunculata are excellent examples of the first kind and included in this same category are the Siphonaptera or fleas, the Hippoboscidae and certain others. Such parasites are insignificant in size compared with their hosts, which are nearly always vertebrates. The hosts have developed a tolerance to their presence and consequently do not suffer fatal effects from their activities. The majority of these parasites are ectoparasites that live on the bodies of their hosts and, since they breathe ordinary atmospheric air, special respiratory adaptations are not required. While the Mallophaga and Siphunculata spend their whole existence on their hosts, in the other groups metamorphosis takes place elsewhere. Excepting the Mallophaga, these parasites are pre-eminently blood-sucking in habit. In adaptation to their special mode of life the skin of such parasites is tough and leathery. In form they are dorsoventrally flattened and are thus enabled to lie close to the host's body; the fleas are exceptional in being laterally compressed. The legs are stout with prominent and often toothed claws which enable them to maintain a firm hold on to their hosts. Eyes are greatly reduced or absent and wings are absent or only rarely present in other than a vestigial condition.

By far the larger number of insect parasites belong to the category of 'parasitoids', which include the Dipterous family Tachinidae, the Hymenoptera Parasitica and several minor groups. They are parasites in the larval condition only and, unlike parasites belonging to the preceding category, the adults into which they transform are active and non-parasitic creatures. They are relatively large in comparison with the size of their hosts, which are almost always destroyed as the result of their activities. The vast majority of hosts are other insects, and these may be parasitized while in the egg or later stages, the imagines, however, being seldom attacked. Among the Tachinidae the parasitism is internal but their eggs may be laid on or away from the hosts. In either event the resultant larvae have to bore their way beneath the skin of their victims: or, they may develop from minute eggs that are adapted to be swallowed by the hosts along with their food. Since no true ovipositor is present in the Tachinidae, only a small proportion of the species are specially adapted for inserting their eggs internally to their hosts. In the parasitic Hymenoptera an ovipositor is universally present; it varies greatly in length and is longest in those species that have to penetrate the wood of trees in order to reach their victims. While many of the parasitic Hymen-optera live as ectoparasites, the greater proportion are endo-parasites and result from eggs inserted directly beneath the integument of their hosts. In their earlier instars, at least, 'parasitoids' feed upon the blood and fat-body of their victims, thus avoiding injury to the more vital organs. Death of the hosts may result from the nutritive drain thus entailed, or the demands of growth may cause the parasites later to turn to other tissues and so to become, virtually, 'internal predators'. The most important adaptations of 'parasitoids' concern respiration. Whereas ecto-

parasites breathe the free atmospheric air and retain in consequence an unmodified peripneustic tracheal system, the endoparasitic forms display very evident adaptations. Whereas many
forms breathe cutaneously the oxygen held in physical solution in
the blood of their hosts, others maintain a more or less direct
connexion with the atmosphere (see p. 87).

Social Insects. Throughout the insects a tendency can be ob-
served for a development of maternal care, usually associated with
some reduction in the number of eggs laid and in the mortality
of the early stages. A social insect may be defined as one in which
parent and offspring live in a common abode or nest. This mode
of existence is made possible owing to the lengthened life of the
female parent which allows of association with her offspring.
Social habits, in this sense, are seen in earwigs, in Embiids, in
certain beetles and in many bees and wasps which are commonly
called solitary. If however we restrict the term social life to those
examples in which an individual cares for offspring which are not
her own, true social behaviour is only displayed in the Isoptera
(p. 163) and among some of the Hymenoptera. In Hymenoptera,
the *workers* are females, the development of whose gonads has
been more or less fully arrested, and though they sometimes lay
some eggs they are never fertilized. Like the *queens* which lay most
of the eggs, they are derived from fertilized eggs, the males or
drones being produced from unfertilized ones.

   In the Vespidae of temperate regions the colonies are annual and
come to an end in autumn, the sole survivors being the young
fertilized females. The workers are not sharply differentiated from
the fertile females, but are distinguishable by their smaller size
and sometimes by certain colour differences. The larvae are fed
by the workers on sugars and masticated fragments of insects.
The nest is either subterranean (*Vespula vulgaris* and *V. germanica*) or arboreal (*Dolichovespula norwegica* and *Vespa crabro*). It
is composed of wood particles mixed with saliva and worked up
by means of the jaws into a substance which, when dry, has a
paper-like consistency. The first formed cells are a small group
suspended by a central pedicel. New cells are added, resulting in
the formation of horizontal tiers or layers, and the whole is enclosed in several layers of covering that protect the colony from
wetness and sudden changes of temperature. The first laid eggs

develop into worker wasps after about four to six weeks from oviposition. These individuals take on the care of the brood and nest, leaving the female parent to devote herself to egg laying. At the end of summer larger cells are constructed and in these develop the females of the next generation. Males develop about the same time, but perish soon after coition, while the females hibernate separately and become the foundresses of the next season's colonies.

Among tropical Vespidae there are wasps that form perennial colonies and a nest may be the combined effort of a number of females and workers. The initiation of new colonies is by a band of females and workers which have issued as a swarm from the parent colony. In temperate regions the colony is founded by a single fertilized female and is said to be *haplometrotic*. It may be argued that if conditions had been favourable the colony would have been perennial and the young females would have remained within the nest, thus giving rise to the *pleometrotic* condition found in the tropics. There is some evidence for this in the gigantic colonies – fourteen feet across – of the *V. germanica* introduced to New Zealand.

The social bees show two major distinctions in their economy when compared with wasps. (1) they often produce wax for comb-building: it is delivered as a secretion of glands located between the plates of the abdominal segments where it hardens into lamellae. (2) They have forsaken a carnivorous diet and have come to use the surer and more easily procurable food supplies afforded by nectar and pollen. Most of the structural peculiarities of bees are consequently adaptations for obtaining these two commodities. Those affecting the mouthparts are alluded to on p. 203. The nectar, it may be added, passes into the crop, where its cane-sugar (sucrose) becomes changed into invert sugar (glucose and fructose). When fed to the larvae it is regurgitated as honey. Pollen adheres to the compound hairs on the body of the worker. It is combed off by means of the spines on the inner side of the first segment of the hind tarsi and then transferred to the corbicula or 'pollen basket'. The latter is either formed by special hairs on the ventral side of the abdomen or, more usually, by the outer surface of the hind legs. The broad tibia and basitarsus are flanked with long hairs for holding the pollen in a compact mass.

In the Bombini or bumble bees the nest is commonly located in some cavity in the ground and its walls are constructed of fragments of grass or moss. It is provisioned with pollen mixed with honey and upon this mass the female constructs a cell of wax within which she deposits her first eggs. The wax, it may be added, is secreted (by the queen and workers) beneath the dorsal and ventral plates of the second to fifth abdominal segments. A waxen receptacle or 'honey-pot' is also constructed and this she fills with honey that functions as a food reserve when she is confined to the nest. After about four days the larvae hatch, and when the stored food is consumed the female opens the cell and regularly supplies the brood with pollen and honey. On the 22nd or 23rd day after oviposition the first workers appear, and when sufficient have emerged they assume the functions of their caste, thus releasing the female for egg laying only. The workers construct further brood cells, together with receptacles for storing honey and pollen. When at full strength a colony numbers from about 100 to 500 bees. The rest of the economy resembles that of *Vespula* and only the young mated females survive the winter. There appears to be no evidence that special food plays any part in queen production, as happens in the honey bee, and the differences between queens and workers are probably attributable to the quantity rather than the quality of the food received. The size difference between these two types of individuals is less constant than in *Vespula* and they tend to intergrade.

As representative of the Apini the honey bee (*Apis mellifera*) shows a higher phase of social economy than is found in the Bombini, and the queens, unlike those of the latter group, are sharply differentiated from the workers. The abdomen, for instance, is more elongated; there is no pollen-collecting apparatus on the hind legs, while both wax and pharyngeal glands are absent. She has become a highly specialized egg laying machine and also produces the 'queen-substance' which the workers lick from her integument; it also inhibits their ovarial development and in its absence they begin building new queen cells. The nest is composed of vertical combs of hexagonal cells placed back to back. The drone cells are larger than those of the workers, while those for queen rearing are irregular and more or less sac-like in form, vertical instead of horizontal, and attached near the edge of the comb. The young larvae are first nourished upon so-called 'royal jelly' that is produced by the pharyngeal glands of worker bees.

Larvae destined to grow into queens are fed with this substance until pupation, whereas with drone and worker larvae it is replaced from the fourth day onwards by a mixture of honey and pre-digested pollen. Experiments of transferring eggs or very young larvae from worker to queen cells, and the reverse, indicate the paramount influence of food in production of these two types of individuals. In this way an egg from a worker cell can be made to develop into a queen and likewise one from a queen cell will grow into a worker bee. The colonies are perennial and, unlike those of bumble bees, are of great numerical strength; a flourishing hive may contain 50 000 to 80 000 bees, of which the vast majority are workers.

All species of ants are social in behaviour, and it is among these insects that caste differentiation reaches its maximum. Thus, in some there are two forms of queen, in others it is the male that is dimorphic: in such cases one form is winged and the other apterous. The workers are wingless with reduced eyes and, where-as these individuals are alike in some primitive groups, in other ants they may be highly polymorphic. In certain species they present a graded series with major workers at one extreme and minor forms at the other. Or, there may be two contrasted types only: large-headed forms with powerful jaws, termed soldiers, that aid in the defence of the colony, and minor forms or true workers. The majority of ants excavate cavity nests in the soil or mound nests composed of various materials; in tropical lands there are species that utilize cavities in plants or construct nests of carton or silk attached to trees. In any case no comb is built, the brood being dispersed in groups along the galleries and chambers of the nest. At times swarms of winged males and females leave the colony. After a brief aerial existence they mate and the females cast off their wings. The young queen forms the beginnings of a new nest wherein she produces the first batch of worker ants. These soon take charge of the colony, thus leaving the parent free to devote herself to oviposition.

According to the theory of W. M. Wheeler, 'trophallaxis', or the mutual exchange of nourishment between adult and young, is usually the bond maintaining social life. Thus, wasp larvae exude saliva in exchange for food supplied by the workers, who eagerly imbibe this secretion: it may even be demanded by the workers without reciprocal exchange of food. While the larvae of some ants likewise exude saliva attractive to their attendants, others produce an integumentary exudate for the same purpose, and in one tropical group the desired secretion is a product of special larval organs. Trophallaxis occurs also among termites but is not found in bees; perhaps the ready availability to bees of pollen and nectar has rendered the exploitation of larval secretions unnecessary. Their function in wasp larvae is also problematic. On Wheeler's theory it would seem that larval secretions, rather than the exercise of a brood care instinct, initiates and sustains the relationship between many social insects and their progeny.

5

Nomenclature, Classification and Biology

Where two kinds of animals differ from each other in some definite but relatively minor structural character or characters, they are said to be of distinct *species*. These structural differences are presumed or, in some cases, known to be, so to speak, indicators of a barrier to interbreeding between the two kinds. Species are grouped into genera, a *genus* being an assemblage of species showing evidence in common characters of close relationship. Genera in their turn are classed into the higher category of a *family*, whose components share many important characters. An *order* comprises all those families that show major features that link them together into a single natural assemblage. To continue, on the ascending scale, orders collectively form a *class*, while classes are grouped on the basis of common fundamental characters into a *phylum*. Certain intermediate grades are also adopted, the most noteworthy being the *subfamily* or group of genera forming a section only of a family; the *superfamily*, or smaller group of families than an order; and the *suborder*. It will suffice to give one example, viz. *Formica rufa* Linnaeus.

SPECIES: *rufa* Linnaeus
GENUS: *Formica* Linnaeus
SUBFAMILY: Formicinae
FAMILY: Formicidae
SUPERFAMILY: Scolioidea
SUBORDER: Apocrita
ORDER: Hymenoptera
CLASS: Insecta
PHYLUM: Arthropoda

As the American taxonomist G. F. Ferris has remarked, classification deals with concepts only. An order, a family or a genus are purely taxonomists' ideas as to how the aggregates termed species may be grouped together so as to show their relationships. The first step in the classification of any group of animals is to have these animals named so as to allow of future reference. The system of naming in universal use is binomial, i.e., each kind of animal bears two names, one generic and the other specific. It dates from the publication of the 10th edition of the *Systema Naturae* of Linnaeus in 1758, when binomial nomenclature was first definitely established for zoology. During the subsequent growth of this science nomenclature has become a matter of great complexity and, in order to regulate the procedure to be followed, a Code of Rules of Nomenclature has been adopted since 1901 by international sanction of zoologists.

The Rules of Nomenclature lay down that scientific names of animals must be Latin or latinized words, or considered and treated as such when not of classical origin. The name of a family is formed by adding the suffix *idae*, and of the subfamily by adding *inae* to the stem of the name of the type genus, e.g., *Blatta*, Blattidae. No rules are laid down for names in higher categories than that of a family. A generic name is a substantive in the nominative singular. The name of a species must be either an adjective agreeing grammatically with the generic name, e.g., *Musca domestica*; or a substantive in apposition with the generic name, e.g., *Stratiomyia chamaeleon*; or substantive in the genitive, e.g., *Psila rosae*. Geographic names and names of persons when used as specific designations are also expressed in the genitive, e.g., *Phormia terraenovae*, *Odontotermes horni*. When the name of the author is quoted after a specific name it follows without mark of punctuation, e.g., *Tabanus rusticus* Linn. When a species is transferred to a genus, other than in that in which it was originally placed, the name of the author of that species is then given in parenthesis, thus: *Blatta lapponica* Linn. has become *Ectobius lapponicus* (Linn.); it will be noted that the specific name conforms grammatically with the new generic name. The valid name of a species or genus is that name under which it was first properly designated and no name published prior to the 10th Edition of the *Systema Naturae* is valid. A generic or specific name that has been replaced on account of its being invalid is known as a *syno-*

nym. It may be also noted that the specimens from which the published descriptions of species are drawn up are called *types*: they are of various categories and have received special names. The importance of types being carefully preserved is obvious when it is realized that a large amount of existing taxonomic work is dependent upon access to them for its ultimate clarification. Descriptions which, at one time, were deemed sufficient so often, in later years, prove inadequate in the light of more exact and discriminating standards. The type is then the final appeal in matters of doubtful identity.

The Classification of Insects. In striving after a natural system of classification of insects, the characters of most importance in diagnosing the main divisions are (1) the presence or absence of wings and their main features, (2) the mouthparts and their changes in ontogeny, (3) metamorphosis and (4) characters afforded by the antennae and tarsi. Orders of most importance to the elementary student are indicated ★.

Group APTERYGOTA

Apterous insects, the wingless condition being primitive, with slight or no metamorphosis; Moulting continues after sexual maturity.

Order 1	Thysanura★	Order 3	Protura
″ 2	Diplura★	″ 4	Collembola

Group PTERYGOTA

Winged insects which are sometimes secondarily apterous; metamorphosis very varied, rarely slight or wanting; Sexually mature adult does not moult.

Division 1. EXOPTERYGOTA (= Hemimetabola)

Insects passing through a simple and sometimes slight metamorphosis, rarely accompanied by a pupal instar. The wings develop externally and the young are generalized nymphs.

Order	5	Ephemeroptera*	Order	13	Dictyoptera*
,,	6	Odonata*	,,	14	Isoptera*
,,	7	Plecoptera	,,	15	Zoraptera
,,	8	Grylloblattodea	,,	16	Psocoptera
,,	9	Orthoptera*	,,	17	Mallophaga
,,	10	Phasmida	,,	18	Siphunculata
,,	11	Dermaptera	,,	19	Hemiptera*
,,	12	Embioptera	,,	20	Thysanoptera

Division 2. ENDOPTERYGOTA (= Holometabola)

Insects passing through a complex metamorphosis always accompanied by a pupal instar. The wings develop internally and the larvae are usually specialized.

Order	21	Neuroptera*	Order	26	Diptera*
,,	22	Coleoptera*	,,	27	Lepidoptera*
,,	23	Strepsiptera	,,	28	Trichoptera*
,,	24	Mecoptera	,,	29	Hymenoptera*
,,	25	Siphonaptera			

ORDER 1. **THYSANURA** (*thusanos*, a fringe; *oura*, a tail)

Small insects usually with compound eyes and often with scales. Antennae long, setaceous and devoid of segmental muscles. Mouthparts for biting; normal and ectognathous. Abdomen with long cerci, 11th segment forming a median tail filament; styli and protrusible vesicles present, ovipositor long. Tarsi 2–4 segmented. Bristle-tails.

These insects mostly occur under stones, logs or among dead leaves, but a few live differently. Thus, *Lepisma saccharina* and some related species, together with *Thermobia domestica*, inhabit dwellings and bakehouses, while *Petrobius maritimus* lives just

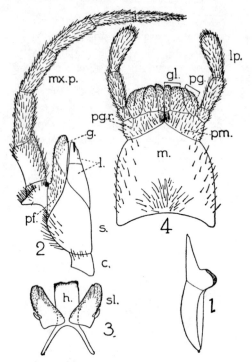

Fig. 60. Mouthparts of *Petrobius maritimus*

1, Mandible. 2, Maxilla: *pf*, palpifer. 3, Hypopharynx (*h*) and superlinguae (sl).
4, Labium: *m*, postmentum. Other lettering as in Fig. 4

above tide mark on rocky coasts; various other Thysanura occur in nests of ants and termites. The mouthparts (Fig. 60) resemble those of Orthoptera and have little in common with their counterparts in Diplura. Genitalia are well developed, but neither the limb-bases nor the styli of the genital segments enter into their formation. Consequently, the long annulate ovipositor is 4-valved only (Fig. 61 D) and there are no claspers in the male (Fig. 61 C). Malpighian tubes are well developed but variable in number, and there are 9 or 10 pairs of spiracles. About 550 species are known and they are grouped into two super families. The Machiloidea are the more primitive and they show certain convergent resemblances to higher Crustacea. The abdominal sterna and limb-bases are separate (Fig. 61), styli are borne on segments 2 to 9 and usually on the coxae of the 2nd and 3rd legs (Fig. 61 A). In the Lepismat-

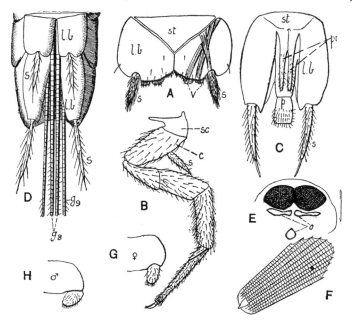

Fig. 61. Thysanura and Diplura

Petrobius: A, 5th abdominal segment, ventral; B, left leg of 3rd pair; C, male genitalia; E, eyes; F, body scale. *Machilis*: D, ovipositor and associated segments (cerci omitted). *Campodea*: G, H, half of 1st abdominal sternum of female and male respectively. *c*, coxa; g_8, g_9, gonapophyses or valves of 8th and 9th segments; *l.b*, limb-base; *o*, ocelli; *p*, penis; *pr*, parameres; *s*, stylus; *st*, sternum; *sc*, subcoxa; *v*, protrusible vesicles

oidea styli are confined to abdominal segments 7 to 9 or 8 and 9 and there are usually no protrusible vesicles.

ORDER 2. **DIPLURA** (*diplos*, double; *oura*, a tail)

Small, eyeless, mostly unpigmented insects with moniliform antennae provided with segmental muscles. Mouthparts for biting, entognathous. Abdomen terminating with variably developed cerci or unjointed forceps: no median tail filament: styli and usually protrusible vesicles present. No ovipositor: tarsi 1-segmented.

About 600 species of Diplura are known. They are denizens of the soil but also frequent decaying vegetable matter of various kinds. The largest forms (Japygidae) measure up to 50 mm long, but an average is 2 to 5 mm in length. Excepting Japygidae, with their darkened sclerotized forceps, these insects are unpigmented. While formerly included in the Thysanura, Diplura are obviously a separate group, perhaps nearer to the ancestral insects. The multiarticulate antennae, provided with segmental muscles, separate the order from most other insects and link it with the Myriapoda. The mouthparts (Fig. 62) are entognathous to the ex-

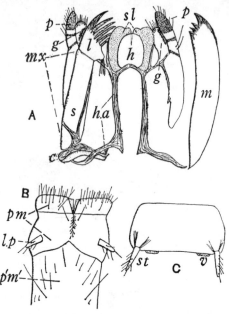

Fig. 62. Diplura

A, *Japyx*, mouthparts. B, labium. C, *Campodea*, 4th abdominal sternum. *c*, cardo;
g, galea; *h*, hypopharynx; *h.a*, hypopharyngeal apodeme; *l*, lacinia; *m*, mandible:
mx, maxilla; *lp*, labial palp; *p*, maxillary palp; *pm*, prementum; *p′m′*, postmentum;
s, stipes; *sl*, superlinguae; *st*, stylus; *v*, protrusible vesicle

tent that the mandibles and maxillae lie in pockets from which they are protruded when feeding. Styli are present on abdominal segments 1 to 7 or 2 to 7, and protrusible vesicles are usual on segments 2 to 7 (Fig. 62 c). External genitalia are very little developed or wanting. Malpighian tubes are represented by papillae or are absent; while only 3 pairs of spiracles are present in Campodeidae, 9 to 11 pairs occur in other families. There are three principal families, the Projapygidae being the most primitive: their short, segmented cerci are traversed by the ducts of special glands. The Japygidae are all forcipate and the Campodeidae (Fig. 63) have long fragile cerci: only the last named insects occur in Britain where they number about a dozen species.

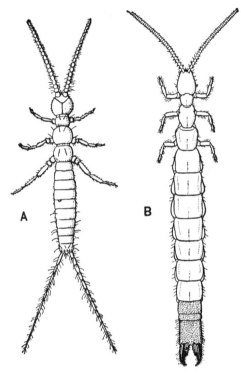

Fig. 63. A, *Campodea.* B, *Japyx*

ORDER 3. **PROTURA** (*protos*, first; *oura*, a tail)

Minute colourless insects without eyes or antennae: mouthparts ento-gnathous for piercing. Abdomen of 11 segments and a definite telson, cerci absent.

These minute creatures seldom attain more than 1 mm in length; they are widely distributed and about 12 species have been found in Britain. They are local and need looking for in moist soil, in turf, beneath bark of trees, under stones, etc. On eclosion from the

egg Protura have eight abdominal segments and three more are
added later, this kind of postembryonic growth or *anamorphosis*
being found in no other insects. Antennal functions are performed by the front legs, which are held forward. Organs,
apparently homologous with the postantennal organs of Collembola, are present one on either side of the head. The mouthparts show basic resemblance to those of some of the Collembola.
Rudiments of limbs are borne on the first three abdominal segments and are evidently homologous with the pair on the first
segment of *Campodea* (Fig. 61 G, H). The legs have 1-segmented
tarsi and single claws. Tracheae are only slightly developed as in
Eosentomon, or are absent as in *Acerentomon*; two pairs of spiracles
occur in the tracheate forms. About 200 species of Protura are
known. The order is, apparently, a divergent offshoot from the
ancestors of insects: nothing is known of the embryology.

ORDER 4. **COLLEMBOLA** (*kola*, glue; *embolo*, a peg)

Very small insects with entognathous, biting mouthparts: antennae
4-segmented: tarsi wanting. Abdomen of 6 segments sometimes fused
together: 1st segment with a sucker-like ventral tube, 4th usually with
a forked springing organ. Spring-tails.

Collembola (Fig. 64) rarely exceed 5 mm long and occur from
the poles to the equator. They are often immensely abundant as
individuals and occur on and below the ground, among herbage, in
decaying matter, under bark, in nests of ants and termites, etc. An
acre of meadow has been found to support nearly 230 000 000 of
these insects from the surface to a depth of nine inches. The eyes
are typically eight ocelli on each side, but may be wanting: just
behind the antenna there is often a characteristic *postantennal*
organ. The mouthparts resemble those of Diplura and, when
feeding, they are partly extruded from pockets within the head;
well-developed superlinguae are present. The antennae, like those
of Diplura, have intrinsic muscles – a feature shared by no other
insects. The abdomen usually bears three ventral structures derived from paired appendages; those of the first segment are
fused to form the *ventral tube*, which is probably an organ for respiration, for absorbing moisture from surfaces, and for adhesion.
A *furca* or springing organ is borne on the fourth segment and,
when not in use, is retained beneath the abdomen by a 'catch' or

retinaculum formed by the reduced appendages of the third segment. External genitalia are wanting and the sexes are almost always alike. Except in *Sminthurus* and its allies trachae are wanting (see p. 78); there are no Malpighian tubules and excretion is performed partly by the fat-body and partly by the periodic shedding of the cells of the mid gut. The gonads have the germarium lateral in position and ovarioles are not developed. Segmentation of the egg is at first total and there is no amnion or

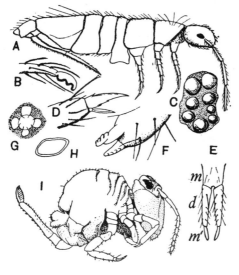

Fig. 64. Collembola

A, *Entomobrya*. B, do., apex of furca. C, eyes of left side. D, apex of tibia with claw. E, *Sminthurus*, furca: *m*, manubrium; *d*, dens; m_1, mucro. F, *Isotoma*, retinaculum. G, *Hypogastrura*, postantennal organ. H, *Isotoma*, postantennal organ. I, *Sminthurides*. (*After* Willem)

serosa. The order is divided into the suborders **Arthropleona** (Fig. 64 A), with cylindrical body and without evident fusion of its segments, and **Symphypleona** (Fig. 64 I) with globular body and with the thoracic and first four abdominal segments closely amalgamated. About 1500 species of Collembola are known: several rank as injurious, the most important being *Sminthurus viridis*, which is a pest of clover and lucerne. Collembola have no close affinity with other insects but it is convenient to leave them in the class.

ORDER 5. **EPHEMEROPTERA** (*ephemeros*, living a day; *pteron*, a wing)

Soft-bodied insects with large eyes, minute antennae and atrophied mouthparts. Wings membranous, longitudinally plicated, hind pair small or even atrophied. Cerci slender, many-jointed, usually accompanied by a median caudal filament. Nymphs aquatic, with plate-like or filamentous tracheal gills. Mayflies.

The Ephemeroptera (Fig. 65) are known as mayflies, many of which live only a few hours as imagines – hence the ordinal name; this feature is compensated by the lengthy nymphal life which may last three years. The venation is very primitive with all the main veins and their branches present: unlike most recent insects the media consists of both MA and MP. The characteristic plication of the wings enables both convex and concave veins to be easily identified. Between the forked branches of the main veins *intercalary veins* are present as in Odonata. While the hind wings are largest in Siphlonuridae, in other families they are much reduced or even atrophied as in *Cloeon* and *Caenis*. The legs are useless for walking and only enable these creatures to cling to objects while resting. The first winged stage is the *subimago* which resembles the imago except for a translucent pellicle which covers the whole insect, giving it a dullish appearance. After undergoing an ecdysis, unique among insects, the true imago is assumed and is recognizable by its clear shining appearance and full coloration. Mayflies take no food as imagines: the alimentary canal remains in a thin-walled condition and is used for taking in air, the mid intestine acting as an aerostatic organ. The nymphs are essentially phytophagous and like the imagines have long cerci and usually a median caudal filament. They inhabit lakes, ponds and streams and present notable adaptive modifications. Burrowing forms (*Ephemera, Hexagenia*) have cylindrical bodies and fossorial fore legs. Species inhabiting swift streams have flattened bodies and hooked spines for clinging to rocks (*Iron, Epeorus*). Inhabitants of sandy streams have the gills covered by opercula formed by the upper lamellae of the second pair: the branchial chamber thus formed is guarded by hair-fringes against the entry of particles suspended in the water (*Caenis, Tricorythus*). Over 40 species of mayflies occur in Great Britain and over 400 kinds are known in North America, while about 2000 species are known in the whole world.

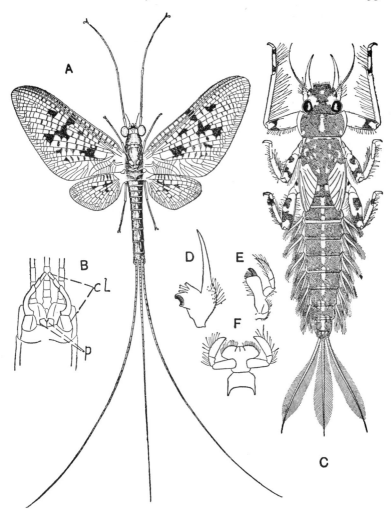

Fig. 65. Ephemeroptera

A, *Ephemera*, male. B, genitalia; *cl*, clasper; *p*, penes. C, *Polymitarcys*, nymph.
D, mandible of same. E, maxilla. F, labium. (*Adapted from* Needham)

ORDER 6. **ODONATA** (*odous*, gen. *odontos*, a tooth)

*Large insects with very elongate bodies, large eyes, and minute
antennae. Mouthparts specialized and strongly toothed: cerci small,
1-segmented. Two pairs of membranous, glassy wings each with a*

pterostigma and numerous cross veins. Nymphs aquatic: labium modified into a retractible prehensile organ. Dragonflies, Damsel-flies.

All Odonata are predators and devour insects of various kinds which they seize while in flight. The capture is effected by the legs, which bear a spiny armature for the purpose on the femora and tibiae. The prothorax is very small, but the meso- and metathorax have conspicuously enlarged pleurites that slant steeply backward. This results in the terga and wings being pushed posteriorly, while the sterna become situated far forward (Fig. 67 A). The legs, in consequence, lie close behind the mouth and are thus enabled readily to seize the prey: they are unfitted for locomotion and the tarsi are 3-segmented. The venation shows no close affinity with that of other insects, and the wings, like those of mayflies, are incapable of being folded with the hind margin over the back. Each wing (Fig. 66 A) bears a pterostigma and vein Sc ends in a conspicuous *nodus* (*n*) or incision, near the middle of the costal margin. The division of the wing area into numerous quad-

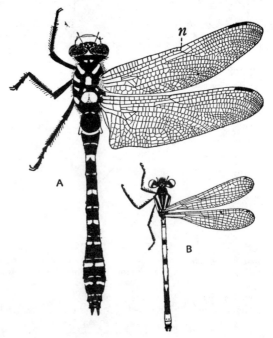

Fig. 66. Odonata
A, Anisopteran dragonfly; *n*, nodus. B, Zygopteran dragonfly

rate cells by a multitude of cross-veins is also a characteristic feature. The abdomen is composed of 10 segments with vestiges of an eleventh segment. The male genital armature is unique in being located on the ventral side of the second segment. The gonopore, however, lies on the ninth segment and the semen is transferred to the penis. During pairing the clasping organs at the apex of the abdomen of the male are used to seize the female by the head or prothorax: the female then curves her abdomen so as to bring the genital opening in contact with the penis. The whole process

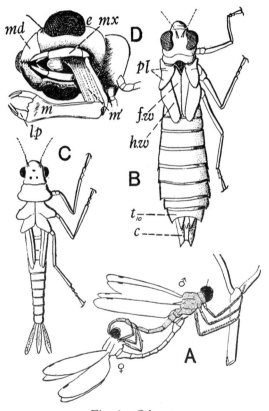

Fig. 67. Odonata

A, position assumed at the end of mating flight in Anisoptera. B, Anisoptera, nymph. C, Zygoptera, nymph. D, ventrolateral view of head of dragonfly nymph with mask (m, m') half retracted. c, cercus; e, compound eye; $f.w$, fore wing; $h.w$, hind wing; $l.p$, labial palp; m, prementum; m', postmentum; md, mandible; mx, maxilla; pl, pleura of meso- and metathorax; t_{10}, 10th abdominal tergum. (A, *after* Meisenheimer)

occurs while the insects are in flight (Fig. 67 A). There are usually marked colour differences between the sexes and the female often has a short 3-valved ovipositor. Oviposition may be *endophytic*, i.e., the eggs are inserted into slits cut by the ovipositor in aquatic plants; or, *exophytic*, i.e., the eggs are dropped freely in the water or superficially attached to vegetation. The nymphs are aquatic and prey upon small Crustacea, insects, etc. They are somewhat sluggish and more or less protectively coloured. Their most characteristic feature is the modification of the labium into a prehensile organ or 'mask' (Fig. 67 D). This organ is hinged between the pre- and postmentum and stowed away between the legs: when a victim is seized the mask is suddenly extended with great rapidity and the prey impaled on the spines of the labial palpi. Dragonflies are divided into two main suborders (Fig. 66), viz. (1) the **Zygoptera,** which have very slender bodies; the two pairs of wings are alike with narrow bases and are held vertically above the abdomen when at rest. (2) The **Anisoptera,** which have stouter bodies, with the hind wings broader basally than the fore wings and the two pairs held horizontally in repose. The nymphs of the Zygoptera (Fig. 67 C) are slender in form and the body is terminated by three elongate caudal gills. In the Anisoptera (Fig. 67 B) the nymphs are of more robust build; caudal gills are absent, but there is an elaborate system of tracheal gills that project in longitudinal rows into the cavity of the hind gut. About 5000 species of dragonflies are known: of these, over 360 species occur in North America and 48 species are found in Great Britain.

ORDER 7. **PLECOPTERA** (*plekein*, to fold; *pteron*, a wing)

Soft-bodied insects usually with long thread-like antennae and cerci: tarsi 3-segmented. Wings membranous, hind pair with a plicated anal lobe. Mouthparts for biting, ligula 4-lobed. Nymphs aquatic with filamentous tufted gills. Stoneflies.

This small order includes some 1700 species, of which 34 occur in Great Britain. They are more closely related to the Orthoptera than to any other living order, but the wings have important venational differences and tegmina are undeveloped: also, there is no ovipositor. Stoneflies are weak fliers seldom found far from the streams inhabited by their nymphs. These are carnivorous or herbivorous and breathe by filamentous gills which are usually disposed in tufts near the bases of the legs, cerci, etc. (Fig. 68), though in the family Eustheniidae of Australia and New Zealand, pairs of lateral gills are situated on the abdominal segments. Nymphal life is often long and, in *Perla*, may last nearly four years: during that time more than 30 instars may be passed through.

ORDER 8. **GRYLLOBLATTODEA** (genus *Grylloblatta*)

Apterous, eyes reduced or absent, no ocelli. Antennae moderately long, filiform. Mouthparts for biting. Legs approximately similar to one another, tarsi 5-segmented. Cerci long, 8-segmented, female with a long ovipositor.

This small order includes 8 species in three genera found principally in the Rocky Mountains, where *Grylloblatta campodeiformis* was discovered by Walker in 1914, but also in Japan and Russia. While showing some specializations, such as the loss of eyes and wings, they appear to be the living remnants of the stock from which both the Orthoptera and Dictyoptera were derived.

ORDER 9. **ORTHOPTERA** (*orthos*, straight; *pteron*, a wing)

Fore wings modified into tegmina. Mouthparts for biting. Hind legs usually modified for jumping, tarsi nearly always 3- or 4-segmented. Female with a well-developed ovipositor, cerci short, nearly always unsegmented. Specialized stridulatory and auditory apparatus often developed. Grasshoppers, Locusts, Crickets.

These insects are sometimes called Saltatoria because they are all more or less capable of jumping. They fall into three principal very large groups. The Long-horned Grasshoppers (Tettigoniidae and their allies) have 4-segmented tarsi, long antennae, and a laterally compressed, often curving ovipositor. The species are often vegetarian and occasionally minor agricultural pests, but others are predatory, often on other Orthoptera.

The Crickets and Mole-crickets (Gryllidae and Gryllotalpidae) have 3-segmented tarsi, long antennae, and long straight (or in mole-crickets vestigial) ovipositors. Crickets are mostly omnivorous and live on or under the ground, but the tree-crickets live on the leaves of bushes. The mole-crickets are subterranean and frequently do serious damage to the roots of crops.

All the preceding groups, apart from a few exceptional mem-

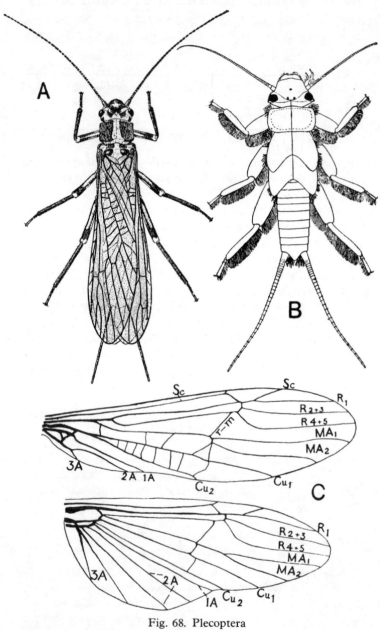

Fig. 68. Plecoptera
A, *Isoperla*. B, a Perlid nymph. C, right wings of *Nemoura*. (A and B *adapted from* Frison)

bers, stridulate in the male by rubbing modified areas of the right and left fore wings together. In both sexes, there is an auditory organ in the fore tibia.

The Short-horned Grasshoppers and Locusts (family Acrididae and its allies) have with few exceptions 3-segmented tarsi, short antennae, and a short stout ovipositor. There are about 9000 species, found predominantly in the warmer regions of the world. Many besides the notorious locusts are serious pests of agriculture. In the majority of species, both sexes but especially the male stridulate by rubbing the hind femur against the fore wing and there is an auditory organ at the base of the abdomen.

The Phases of Locusts. Locusts are grasshoppers of not necessarily closely related genera, the individuals of which when crowded tend to cohere in bands or swarms.

Each species of locust has the capacity to develop into forms or *phases* that differ from one another both structurally and biologically. At one extreme these insects may occur in the migratory phase (*phasis gregaria*), when they are highly destructive to vegetation: at the other extreme they may be in the solitary phase (*phasis solitaria*), when they behave after the manner of ordinary grasshoppers and are relatively harmless. Formerly these two phases were sometimes considered to represent separate species, but are now known to be modifiable and to be connected by intermediate forms that constitute the *phasis transiens*.

Among the best known species are the Migratory Locust (*Locusta migratoria*) of the Old World; the Desert Locust (*Schistocerca gregaria*) of Africa and Western Asia, and the Brown Locust (*Locustana pardalina*) of South Africa. *Locusta migratoria* may be taken as an example. In the phasis gregaria it is characterized by the black and orange nymphal coloration, that develops independently of the nature of the environment. The adults have the pronotum laterally constricted and without a dorsal carina; the hind femora are relatively shorter and the tegmina longer than those of individuals in the solitary phase. In the phasis solitaria the nymphal coloration is plastic with a marked tendency to simulate that of the immediate environment. In the adult the pronotum is longer, it is not laterally constricted and bears a dorsal carina (Fig. 69).

The phase theory has been subjected to experiment, and given suitable temperature, humidity and food it is possible to control

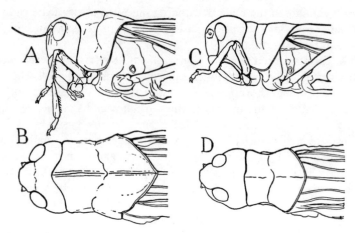

Fig. 69. Form of thorax in phases of the locust (*Locusta migratoria migratorioides*)
A, B, phasis solitaria. C, D, phasis gregaria. (*After* Faure)

the phase that develops. Individuals reared from the egg in separate containers assume evident solitaria characters, whereas those reared collectively in large numbers show strong gregaria features; under less crowded conditions they develop into the phasis transiens. The typical or ancestral form of the species is displayed by the solitary phase and the differences shown by individuals in the migratory phase are perhaps due to their great activity. It is probable that the characteristic black and orange nymphal pigmentation result as a by-product of a high rate of metabolism. Out in the field the migratory phase only develops in specific breeding areas where a particular combination of ecological factors prevail. For this reason the destruction or modification of these areas will eliminate the conditions favouring the development of this phase and thereby open the way for countering the tendency to form migratory swarms. The latter, it may be added, do not arise as the result of food shortage, but appear to be connected with the maturation of the gonads and the wind-assisted dispersal of the species. The resultant progeny of these swarms develop into solitaria or transiens individuals should their environment be different from the original breeding grounds.

ORDER 10. **PHASMIDA** (*phasma*, an apparition)

Large, often apterous insects, frequently of elongate, cylindrical form, more rarely depressed and leaf-like. Mouthparts for biting. Prothorax short, legs similar to one another, tarsi nearly always 5-segmented. Cerci short, unsegmented: ovipositor concealed. Stick- and Leaf-insects (Fig. 70 A).

This order includes about 2,500 predominantly tropical species. Most are sluggish and resemble in form and colour some part of the plant on which they feed. Males are often rare and the eggs, which are laid singly and usually fall to the ground, often develop parthenogenetically. None of them possess a specialized stridulatory or auditory apparatus.

ORDER 11. **DERMAPTERA** (*derma*, skin; *pteron*, a wing)

Fore wings represented by small tegmina: hind wings large, membranous and complexly folded. Mouthparts for biting, ligula 2-lobed: body terminated by forceps. Earwigs.

This order numbers about 1200 species. *Forficula auricularia* is the common Earwig of Europe which has become established in North America and New Zealand. It is omnivorous in habit and the female shows parental care for the eggs and young nymphs. The forceps are modified cerci and, in some forms, are present as many-jointed appendages during the immature stages. *Hemimerus* and *Arixenia* are aberrant; the former is a wingless ectoparasite of the rat *Cricetomys*; the latter, also wingless, lives in caves with bats.

ORDER 12. **EMBIOPTERA** (genus, *Embia*; *pteron*, a wing)

Elongated soft-bodied insects living in silken web-like tunnels. Two pairs of equal-sized long, smoky wings usually with signs of venational degeneration. Mouthparts for biting, ligula 4-lobed: tarsi 3-segmented, 1st segment of anterior pair greatly swollen: cerci 2-segmented. Females apterous. Web-spinners.

The members of this small order live in silken tunnels beneath stones or loose bark of trees. They are gregarious and several individuals, along with nymphs and eggs, inhabit one tunnel system. The silk is produced by numerous glands lodged in the inflated first segment of the anterior tarsi: the ducteole from each gland discharges ventrally at the apex of a corresponding bristle. These

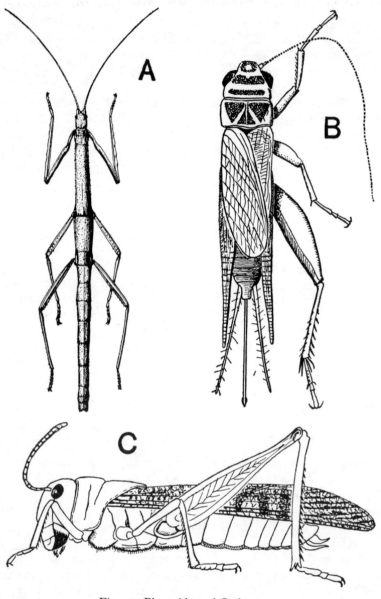

Fig. 70. Phasmida and Orthoptera

A, Stick-insect (*Carausius*), female (Phasmidae). B, Cricket, female (Gryllidae).
C, Locust, female (Acrididae)

glands are present in individuals of both sexes and of all ages. Except in the primitive genus *Clothoda*, the tenth abdominal segment and the cerci are modified asymmetrically in the male for pairing. About 300 species are known and they occur throughout the warmer regions of the world. One or two species only are found in South Europe and fewer than 20 inhabit the warmer parts of the United States. The Embioptera are rather an isolated order and a form not unlike *Clothoda* is known from the Lower Permian.

ORDER 13. **DICTYOPTERA** (*dictyon*, a network; *pteron*, a wing)

Antennae nearly always filiform with many segments. Mouthparts for biting. Legs similar to each other or fore legs raptorial, tarsi with 5 segments. Fore wings more or less thickened into tegmina with marginal costal vein. Cerci many-segmented, ovipositor reduced and concealed, eggs contained in an ootheca. Cockroaches, Mantids.

Although often put in one order with Grasshoppers and Stick-insects, they are in many ways distinct, as indicated above. The Cockroaches or Blattaria, include about 4000 species, predominantly tropical in distribution and typically nocturnal and omnivorous or vegetarian. The eggs are enclosed in an ootheca which is carried about by the female for a longer or shorter period.

The Mantodea include about 2000 species which are exclusively carnivorous and are found in all the warmer parts of the world. They capture other insects by means of their front legs which have rows of spines on the apposable femur and tibia (Fig. 5).

ORDER 14. **ISOPTERA** (*isos*, equal; *pteron*, a wing)

Social insects living in large communities: soft-bodied and generally pale coloured. Mouthparts for biting, ligula 4-lobed: cerci very short. Either with two pairs of elongated similar wings, which are soon shed, or without wings. Apterous forms with rudimentary eyes or none and mainly of two types, viz., soldiers with large heads and jaws or a pointed rostrum and workers with normal heads and jaws. Termites or White Ants.

The members of this order are, structurally, closely allied to the Blattaria. All termites are social and polymorphic and live as highly organized colonies in nests or *termitaria*. The primitive species merely tunnel into wood, but others form special nests,

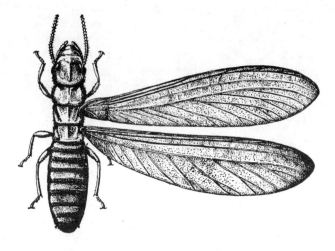

Fig. 71. Macropterous form of *Kalotermes*
(*From* Banks and Snyder)

often of great size, and composed of earth and wood mixed with faecal matter and saliva. Four chief *castes*, composed of individuals of both sexes, may occur in a species. These castes comprise functional reproductive forms of two kinds and sterile forms of two principal kinds. The usual reproductive caste is the fully-winged or *macropterous* form (Fig. 71). At suitable times individuals of this caste swarm from the nest, cast their wings and, after mating, found new colonies. Brachypterous and apterous forms may also occur and are, functionally, the supplementary reproductive caste. In the more specialized termites only a single fertilized female is present in a colony. This individual, which may be derived from either of the two castes just named, undergoes remarkable post-metamorphic growth and may be as long as $4\frac{1}{2}$ inches. In the sterile castes the gonads are greatly reduced and non-functional in both sexes. The members are divisible into *workers* and *soldiers* (Fig. 72). The *workers* are the most numerous individuals in a colony: they build the nest and keep it provisioned. Owing to their gnawing propensities they have earned for the termites their

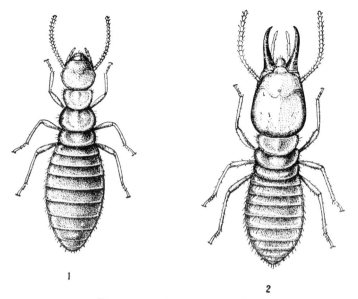

Fig. 72. *Prorhinotermes simplex*
1, worker; 2, soldier. (*After* Banks and Snyder, *U.S. Nat. Mus. Bull.*, 108)

notoriety as destroyers of woodwork and other materials. The *soldiers* act as defenders of the nest and usually have large sclerotized heads and jaws. In the specialized genus *Nasutitermes* they are replaced by *nasute forms* with pyriform heads drawn out into a rostrum, and very small jaws. A *frontal gland* discharging a defensive secretion through a *frontal pore* is commonly present. This type of chemical warfare is specially developed in the nasute forms, which emit a pungent secretion through the rostrum. The lower termites feed on wood and harbour a rich Protozoan fauna in the hind intestine. The most characteristic of these organisms belong to the groups Polymastigina and Hypermastigina, whose only known host, other than termites, is the wood-feeding cockroach *Cryptocercus* of North America. Functionally, the Protozoa are symbionts which produce enzymes that split up the cellulose of the wood into assimilable products: termites sterilized of their Protozoa are unable to digest wood. The higher termites feed on fungi, humus and organic matter from the soil. About 1900 species of Isoptera are known. The few European forms are found in southern parts of that continent: about 55 species occur in North America, but most termites are denizens of the tropics.

ORDER 15. **ZORAPTERA** (*zoros*, pure; *apteros*, wingless)

Minute winged or apterous insects. Mouthparts for biting. Thorax with the segments of nearly equal size, tarsi with 2 segments. Short segmented cerci present, ovipositor absent.

The first species of this group were recognized by Silvestri in 1913 and there are now 22 species, all placed in the genus *Zorotypus* and found in all the warmer regions of the world. They live gregariously under bark or in humus. They have been classified as Psocids but the structure of the head and thorax suggest rather Orthopteroid affinities.

ORDER 16. **PSOCOPTERA** (genus *Psocus*; *pteron*, a wing)

Very small soft-bodied winged or apterous insects with modified biting mouthparts. Venation reduced and seldom with cross-veins: fore wings with pterostigma. Tarsi 2- or 3-segmented, cerci absent. Book-lice and allies.

The small insects of this order are world-wide in distribution and nearly 2000 species are known. They have long multiarticulate filiform antennae, the maxillae are single-lobed, each ensheathing an elongate rod or 'pick', and there are no cerci. These insects are known as book-lice, since several kinds often occur among books, etc., in little-used rooms: others are found among stores of

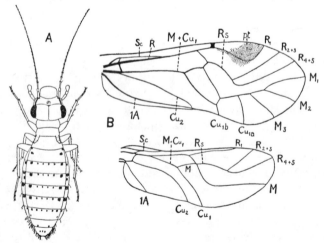

Fig. 73. Psocoptera

A, *Trogium pulsatorium* (*from* Tillyard). B, *Amphigerontia*, right wings cereal products, or of straw or chaff, but more usual habitats are among vegetation or on the bark of trees, lichens, old palings, etc. They are more or less gregarious and lay their eggs in groups covered with silken threads. The rod-like 'pick' represents the lacinia and is possibly used for rasping off fragments of bark or other plant tissues.

ORDER 17. **MALLOPHAGA** (*mallos*, hair; *phagein*, to eat)

Apterous insects living in all stages as ectoparasites of birds or less frequently mammals. Mouthparts of a modified biting type. Prothorax free, meso- and metathorax often imperfectly separated. Tarsi of 1 or 2 segments, with 1 or 2 claws. Cerci absent. Metamorphosis slight. Bird or Biting Lice.

About 2800 species of these creatures are known. They fasten their eggs to the feathers or hair of a warm-blooded host on which the whole life history is spent. If removed from the host, they soon die. Lice from closely related hosts form well-defined groups that are themselves of close relationship, thus suggesting that, to a considerable extent, host and parasite evolution has taken place simultaneously. It is possible that Mallophaga are derivatives of original nest-inhabiting Psocoptera that came to live an ectoparasitic life. Among members of the group the chicken 'mite' *Menopon pallidum* is notorious: species of *Trichodectes* live on dogs, cats and other domesticated animals.

ORDER 18. **SIPHUNCULATA** (*siphunculus*, a little tube)

Apterous insects living in all stages as ectoparasites of mammals. Mouthparts highly modified for sucking and piercing, retracted when not in use. Thoracic segments fused, tarsi 1-segmented, claws single. Cerci absent. Metamorphosis slight. Sucking Lice.

The Sucking Lice, of which about 300 species are known, are often classified with the Biting Lice, but besides having very different mouthparts, the thoracic spiracles are dorsal not ventral. All species are blood-sucking parasites of mammals. The highly modified mouthparts are used for piercing and in front of the head is a small retractile tube or haustellum armed with denticles. When everted the denticles anchor the insect to the skin of the host. Three stylets, said to be derived from the labium and hypopharynx, pass through the haustellum and pierce the skin. These stylets are enclosed in a sheath-like pharyngeal tube which is inserted into the puncture and the action of the cibarial and pharyngeal muscles pumps blood into the gut of the louse. The original mandibles and maxillae on this view are atrophied. *Pediculus* includes species found on man and the higher apes. The best known is *P. humanus* of man, with its subspecies, known respectively as the Head Louse (*capitis*) and the Body Louse (*humanus* s. str.). The human louse is the vector which transmits from man to man the pathogenic agents of epidemic typhus, relapsing fever and trench fever. *Pthirus pubis* is the Crab Louse of man, while *Haematopinus* has common species infesting the pig, horse and ox.

ORDER 19. **HEMIPTERA** (*hemi*, half; *pteron*, a wing)

Wings very variably developed with reduced or greatly reduced venation: Fore pair often more or less corneous: apterous forms frequent. Mouthparts for piercing and sucking with mandibles and maxillae stylet-like and lying in the projecting grooved labium: palpi never evident. An incipient pupal instar sometimes present. Plant-bugs, Cicadas, Leafhoppers, Aphids, Scale Insects.

This order includes more than 56,000 species and is the largest among Exopterygota. Its members are best recognized, in cases of doubt, by the mouthparts, which are very constant in their essential structure. The wings present no common venational or other features and, furthermore, are often absent. The mouthparts are adapted for piercing plant tissues and extracting the sap. The labium or rostrum projects downward from the head and forms a grooved channel within which lie two pairs of needle-like stylets. At the base of the labium the groove is absent and the stylets are roofed over in this region by the labrum. The mandibles form the outer (or anterior) pair of stylets and the maxillae are lodged between. The maxillary stylets interlock so as to enclose an anterior or suction canal and a posterior or salivary canal (Fig. 74). Just before the maxillae diverge in the head a narrow *food meatus* enters the suction canal. The common salivary duct opens into a salivary pump which discharges the salivary secretion at the apex of the hypopharynx into the salivary canal. The mandibles are the chief piercing organs and the maxillae are afterwards inserted into the puncture. Ordinarily, the stylets are forced into the tissues by their protractor muscles, being guided by the labrum and the grooved labium. As they gradually penetrate,

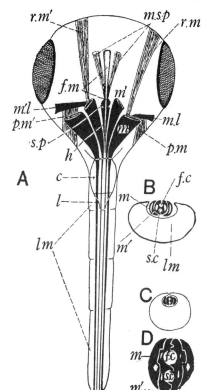

Fig. 74. Hemiptera
A, simplified diagram showing feeding mechanism of a plant bug (Heteroptera), anterior view. B, transverse section of mouthparts, just in front of labrum. C, transverse section of mouthparts near apex of labium. D, transverse section of mouth-stylets of *Anasa tristis* (Heteroptera), showing method of interlocking (*from* D. G. Tower, 1914). *c*, clypeus; *f.c*, suction canal; *h*, hypopharynx; *m.l*, mandibular lever; *m'.l*, maxillary lever; *m.s.p*, muscles of salivary pump; *p.m* protractor muscle of mandible; *p.m'*, do, of maxilla; *s.c*, salivary canal.
Other lettering as in Fig. 75

bringing the head nearer the leaf surface, the labium becomes shortened by being bent or looped (Fig. 75 D). In the scale-insects and white-flies the stylets may exceed the length of the whole insect and when retracted, are coiled within an internal pouch or *crumena* (Fig. 75 E). The four stylets are interlocked to function as a single structure: they are inserted into the plant partly by the action of the protractor muscles and partly by means of a muscular clamp, near the apex of the labium, which alternately grips and releases the stylets after the manner of forceps. Hemiptera feed by drawing the sap into the food meatus by the action of a muscular cibarial pump: it then enters the pharynx and passes into the mid intestine. By virtue of their universal sucking propensities many Hemiptera cause a vast amount of direct or indirect injury to cultivated plants. Certain aphids and leafhoppers also convey highly destructive virus diseases from plant to plant during feeding.

Especially noteworthy are diseases of potatoes, tobacco, maize and sugar-cane that are transmitted in this manner.

Hemiptera are divided into two main groups that are often regarded as separate orders, viz. the Heteroptera and Homoptera. The **Heteroptera** or plant bugs have the fore wings usually modified into hemelytra (Fig. 75 C) and, while at rest, they overlap flat on the abdomen: also the base of labium is separated from the anterior coxae by a sclerotized area of the head wall. The majority of the members of this suborder are plant feeders and among them are a number of injurious species. These include the Chinch-bug (*Blissus leucopterus*) of the United States, the Cotton Stainers (*Dysdercus*) of the tropics and the Apple Capsid (*Plesiocoris rugicollis*) of Europe (Fig. 75 A). A propensity for animal food has been acquired in the predacious family Reduviidae, and most aquatic bugs, which imbibe the body fluids of insects and other small animals. Bed-bugs (*Cimex*) and the Reduviid *Triatoma* comprise blood-suckers of man. A number of Heteroptera are aquatic in habit and include the Water Boatmen (Notonectidae) (Fig. 75 F), Water Scorpions (Nepidae), Giant Water-bugs (Belostomatidae), and others, which are familiar examples.

The **Homoptera** have the fore wings either leathery or membranous but of uniform consistency, and they are folded roof-like along the sides of the body. The labium arises far back on the head or even between the fore coxae. Associated with Homoptera are three characteristic features in their economy, viz. the almost continuous discharge of a sugary waste product or 'honey dew' from the anus, especially notable in aphids; the prevalent excretion of wax either in a powdery form or as threads; and the presence in the abdomen of a peculiar tissue – the *mycetome* – which harbours micro-organisms supposedly symbiotic in function. Among the different families of Homoptera the Cicadas (Cicadidae) are well known for the shrill sounds emitted by the males, and the Lantern-flies (Fulgoridae) are large tropical insects usually of brilliant coloration. The Froghoppers (Cercopidae) are small insects whose nymphs live within a frothy exudation which is believed to prevent desiccation and also to afford protection from enemies. The abdominal spiracles open into a ventral cavity formed by the downgrowth of the tergites and pleurites which meet beneath the sterna. This cavity is closed anteriorly, but air enters posteriorly through a kind of valve. The actual frothing

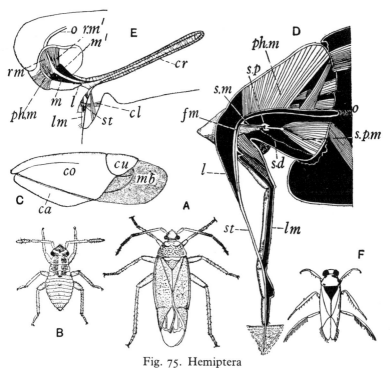

Fig. 75. Hemiptera

A, *Plesiocoris rugicollis* (Miridae). B, 1st instar nymph of same. C, hemelytron of a Mirid. D, section through the head of *Graphosoma* (Heteroptera), showing position of labium during feeding. E, *Pseudococcus* (Coccoidea), lateral view of head showing position of stylets in crumena. F, *Notonecta*. (A and B *from* Petherbridge and Husain: D and E *adapted from* Weber, 1930.) *ca*, clavus; *cl*, clamp muscle; *co*, corium; *cr*, crumena; *cu*, cuneus; *f.m*, food meatus; *l*, labrum; *lm*, labium; *m*, mandibular stylet; *m'*, maxillary stylet; *mb*, membrane; *o*, oesophagus; *ph.m*, dilator muscles of sucking pump; *rm*, retractor muscle of mandible; *rm'*, do. of maxilla; *s.d*, salivary duct; *s.m*, salivary meatus; *s.p*, salivary pump; *s.p.m*, dilator muscles of salivary pump; *st*, stylets

results from the fluid, exuded from the anus, forming a film over this valve and becoming blown into bubbles by air expelled through the latter. The Cicadellidae are leafhoppers and the Aleyrodidae are the white-flies whose bodies and wings are dusted with a powdery wax. The final nymphal instar in Aleyrodidae undergoes a peculiar transformation, recalling the pupal changes in Endopterygotes.

The Aphidoidea (Fig. 76) are an immense group, including many destructive species. They all pass through a more or less

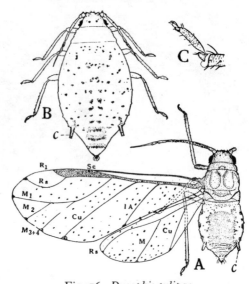

Fig. 76. *Dysaphis tulipae*

A, winged viviparous female. B, wingless do. C, tarsus of 3rd leg. *c*, cornicle
(*Adapted from* J. Davidson)

complex life-cycle on one or more plant hosts and are commonly
termed 'green-fly' or plant-lice. The Aphididae comprise the
greater number of the species: they bear a pair of short dorsal
tubes or cornicles on the abdomen through which an alarm
pheromone can be exuded. In this family the parthenogenetic
generations are viviparous. Winter is usually spent in the egg,
which is commonly laid on a woody plant termed the *primary host*.
In the spring an apterous agamic female (*fundatrix*) hatches out
which produces very similar offspring (*fundatrigeniae*), but soon
winged agamic females (*migrantes*) develop and these fly to some
herbaceous plant or *secondary host*. All through the summer new
generations of winged or wingless agamic females (*alienicolae*) are
produced on the secondary host, the winged forms relieving over-
crowding by flying to other plants. In early autumn winged alienicolae known as *sexuparae* fly to the primary host and produce
males and females (*sexuales*). After mating, eggs are laid and the
cycle is thus completed. Members of the family Phylloxeridae
undergo a more complex life cycle and differ from the Aphididae in that the cornicles are absent and reproduction is never
viviparous. The genus *Phylloxera* (= *Viteus*) is an important

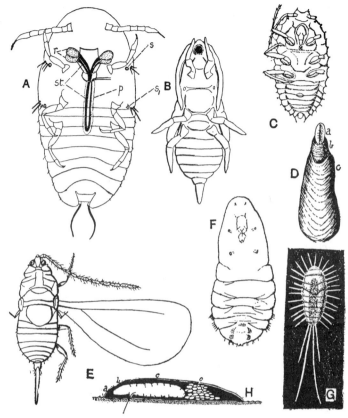

Fig. 77. Coccoidea

A, 1st instar of *Coccus hesperidum*, ventral, × 110. B, *Lepidosaphes beckii* (= *fulva*) male pupa, ventral, × 48. C, *Pseudococcus longispinus* young adult female, ventral, × 28. D. *Lepidosaphes ulmi* female scale, dorsal, × 16. E, *Aspidiotus limonii* male, × 48. F, *Lepidosaphes ulmi* female removed from scale, ventral, × 37. G, *Pseudococcus longispinus* female in natural state on a leaf, × 4. H, *Lepidosaphes ulmi*, longitudinal section of female, × 16. (Figs. D and H original; remainder *adapted from* Berlese.) *a*, scale of first instar; *b*, do. of second instar; *c*, do. of adult; *o*, eggs; *p*, crumena; *s*, *s₁*, spiracles; *st*, mouth stylet

enemy of the vine. The Adelgidae affect Coniferae. Their sexual generation always occurs on spruce (*Picea*) and the agamic forms on *Larix, Pinus, Abies*, etc. The Coccoidea (Fig. 77) include the scale-insects and mealy-bugs. They rank among the most highly modified of all insects and are notable for their extreme sexual dimorphism. In the mealy-bugs the insect is covered with a fine waxy exudation, while in the true scale insects the 'scale' is a covering formed by the persistent exuviae of the previous instars glued together with a dermal secretion. Many coccids are well known injurious insects, and the more important include *Lepidosaphes ulmi*, the Mussel Scale (Fig. 77 D); *Quadraspidiotus perniciosus*, the San Jose Scale; *Icerya purchasi*, the Fluted Scale; *Planococcus citri*, the Citrus Mealy-Bug, and many others. On the other hand, *Dactylopius coccus* yields cochineal, and *Laccifer lacca* of India produces a resinous exudation or lac providing commercial shellac. In the early instars the two sexes of coccids are indistinguishable and the young insects are active with well-developed antennae, legs and mouthparts. Those producing females pass through one or two instars fewer than the males. In the more primitive types, or mealy-bugs, including *Monophlebus*, *Pseudococcus*, etc., the females continue active (Fig. 77 C), but in most forms they become sedentary after the first or second instar and various degrees of reduction of the appendages supervene (Fig. 77 F). Culmination is reached in *Physokermes* and other genera where the antennae and legs have totally atrophied. In the males the third or fourth instar is a prepupa and the succeeding instar is a pupa: during these phases the original appendages disappear and the imaginal organs that replace them are external growths. In the male imago anterior wings only are present, the hind pair being represented by slender halteres that are linked to the wing-bases by hooklets (Fig. 77 E): mouthparts are absent.

ORDER 20. **THYSANOPTERA** (*thusanos*, a fringe; *pteron*, a wing)

Minute slender insects with short 6- to 10-segmented antennae and very narrow wings with long hair fringes. Mouthparts stylet-like, for rasping. Tarsi very short, ending in a vesicle: cerci absent: an incipient pupal stage present. Thrips.

Thysanoptera frequent many kinds of plants besides being found in decaying wood, fungi, etc. About 5000 species are known and they occur all over the world. While rarely exceeding 4 mm long, these insects are very abundant as individuals. Most being sap-feeders, they are often enemies of cereals and other cultivated crops such as peas and fruit trees: their attacks on the flowers often lead to sterility or to the fall of the fruit. The head bears a ventral cone or short rostrum that is formed by the labrum

above, the labium below and laterally by the plates of the maxillae. The rostrum thus encloses the mandible, the two maxillae
and the hypopharynx. The mandible, which is a stout stylet, is the
left one, its counterpart being vestigial. Each maxilla consists of a
palpus-bearing plate, together with a slender 2-segmented stylet
articulating with it. During feeding the rostrum is closely applied
to a leaf; the tissues are lacerated and broken up by the three
stylets named and then sucked up through the rostrum by the
pumping action of the cibarial sucking pump. During growth four
instars commonly occur, the third and fourth instars being the prepupa and pupa, respectively. Both these instars are resting
stages in which no food is taken, and they commonly occur in the
soil. External buds of the wings and appendages are evident, and
these instars have analogies with the holometabolous condition of
the higher insects. While Thysanoptera form an isolated order
they are clearly members of the Hemipteroid group.

ORDER 21. **NEUROPTERA** (*neuron*, a nerve; *pteron*, a
wing)

*Small to large soft-bodied insects with two pairs of membranous wings
without anal lobes: venation generally with many accessory branches
and numerous costal veinlets: Rs usually pectinately branched.
Mouthparts for biting, antennae well developed, cerci absent. Larvae
campodeiform with biting or suctorial mouthparts: predacious and
aquatic or terrestrial.* Alder flies, Lacewings, Ant-lion flies, etc.

This rather heterogeneous order (Fig. 78) is divided into the
suborders Megaloptera and Planipennia. The **Megaloptera** have
the more primitive venation with fewer accessory veins and the
pectination of Rs is usually undeveloped: their larvae have biting
mouthparts. The Alder-flies (*Sialis*) and allies have aquatic larvae
with seven or eight pairs of hair-fringed abdominal appendages.
Corydalis, which occurs in North and South America, attains a
wing expanse of 15 cm., with gigantic mandibles in the male.
The Snake-flies (*Raphidia*) have a long neck-like prothorax and an
elongate ovipositor: their larvae live under bark of conifers, etc.

The **Planipennia** include the majority of Neuroptera. The venation shows a pectinate Rs and a great development of secondary
branching, especially as bifurcations along the wing margins. The
larvae occur on vegetation, or in the earth, or are aquatic. All have
exserted, piercing mouthparts of similar basic design. The man-

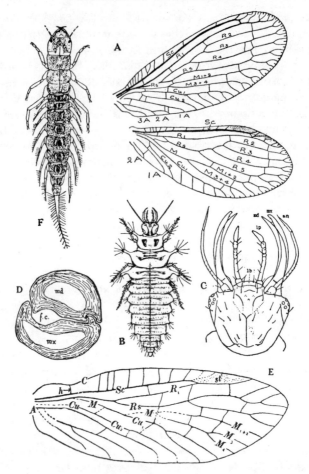

Fig. 78. Neuroptera

A, right wings of a Neuropteron (*Sisyra*) (*after* Comstock). B, Larva of *Chrysopa ciliata* (= *alba*) (*after* Withycombe). C, do., head (*after* Withycombe). *an*, antenna; *lb*, labrum; *md*, mandible; *mx*, maxilla. D, transverse section showing interlocking of mandible and maxilla of *Chrysopa* larva forming a food-channel (*f.c*). E, venation of fore wing of *Sialis* (*after* Needham). F, larva of *Sialis* (*after* Needham)

dibles and maxillae (Fig. 78 C, D) are co-ordinated and enclose a groove-like suction canal through which the body fluids of the prey are imbibed: also, six out of the eight Malpighian tubes become silk glands and they weave their cocoons through an anal spinneret. As in the Megaloptera the pupae are primitive and the pharate adult is capable of walking or climbing before emergence. Of the main families the Hemerobiidae (Brown Lacewings) and Chrysopidae (Green Lacewings) are notably beneficial since their larvae destroy large numbers of aphids and other small insects. The Sisyridae and Osmylidae have aquatic larvae: those of the first named live in association with fresh water sponges, piercing the tissues with their mouthparts. The Nemopteridae have very elongate filiform hind wings: their larvae have very long necks, sometimes exceeding the whole of the rest of their bodies. The Myrmeleontidae are large insects with subfiliform antennae: their larvae, or ant-lions, commonly make pit-like snares for capturing their prey. The Ascalaphidae are closely related, but the antennae are longer and clubbed; their larvae lurk under stones, on leaves or on trees. The Mantispidae are predators and resemble the Mantids in the formation of their raptorial legs: the life cycle involves hypermetamorphosis and the larvae become external parasites of spiders' eggs or of wasps. The Coniopterygidae seldom exceed 8 mm in wing expanse and have greatly simplified venation. Their claim for inclusion in the order is based on the morphological characters of their larvae. They occur on trees and shrubs, the larvae preying upon minute insects, mites, etc.; the adults are covered with a powdery exudation from epidermal glands and resemble white-fly. About 4000 species of Neuroptera are known, and of these 60 occur in Britain.

ORDER 22. **COLEOPTERA** (*koleos*, a sheath; *pteron*, a wing)

Minute to large insects whose fore wings are modified into elytra that meet in a line down the back. Hind wings membranous, folded beneath the elytra or absent, prothorax large: mouthparts for biting. Larvae of diverse types but never typical polypod. Beetles.

The Coleoptera with over 330 000 described species ranks as the largest order in the animal kingdom. While their habits are very varied they are more especially ground insects that live either in the soil, or in decaying matter associated with it. Several families are aquatic and great numbers of species are phytophagous both as larvae and adults. In addition, various species live in timber and dry, stored products. Beetles are very uniform in external structure. The head (Fig. 79) is characterized by the very

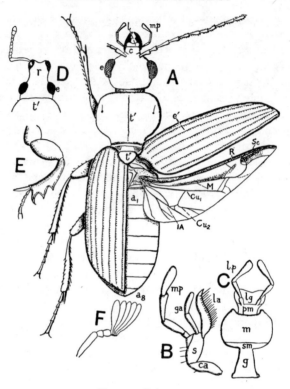

Fig. 79. Coleoptera

A, a beetle, fam. Carabidae. B and C, maxilla and labium of same. D, head of weevil, showing rostrum, r. E, leg of a Scarabaeid beetle. F, antenna of a cockchafer. a_1, 1st abdominal segment; a_8, 8th do.; c, clypeus; ca, cardo; e, compound eye; e′, elytron; g, gula; ga, galea; l, labrum; la, lacinia; l.p, labial palp; lg, ligula; m, mentum; m.p, maxillary palp; pm, prementum; s, stipes; sm, submentum; t′, prothorax; t′′, scutellum

general presence of a *gula* (p. 21) and the legs are well adapted for running or often for burrowing also. The hind wings are the functional organs of flight and, when the insect is in the air, the elytra play no active part in propulsion and are held at an acute angle with the body. The wings are often long and are intricately folded beneath the elytra; sometimes they are much reduced or wanting and, in such cases, the elytra are often soldered together. In the Staphylinidae and related families the elytra are much shortened, while in the Oil Beetles (*Meloe*) they are vestigial. The wing venation is difficult to homologize with that of other insects:

there is a predominance of longitudinal veins and, except in Adephaga, cross-veins are either absent or very few.

Coleopterous larvae (Fig. 80) afford excellent examples of adaptation to particular modes of life. Thus, in the Adephaga and the Staphylinidae the campodeiform type prevails. Such larvae have the antennae, legs and sensoria well developed, thereby fitting them for an active predatory life. Elongate simple, or jointed, processes resembling cerci are borne on the ninth abdominal

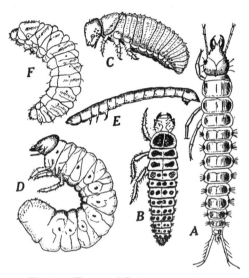

Fig. 80. Types of Coleopterous larvae

A, Ground Beetle (Carabidae). B, Ladybird (Coccinellidae). C, Leaf-beetle (Chysomelidae). D, Chafer (Scarabaeidae). E, Wireworm (Elateridae). F, Weevil (Curculionidae)

segment: the tenth segment usually functions as a pseudopod. Less active larvae, which have no need to seek far for sustenance, are characteristic of a large number of families. They are of modified campodeiform type with the appendages and sensory organs reduced. Examples are larvae of the Coccinellidae (Ladybirds), Elateridae (Wireworms) and Chrysomelidae (Leaf-beetles). The scarabaeoid type of larvae is subterranean in habit and prevails in cockchafers, dorbeetles and their allies. Crescentic in form, it has a large sclerotized head, well-developed legs, a soft, inflated abdomen and conspicuous cribriform spiracles. Among wood-boring larvae the body is usually fleshy and unpigmented with stout jaws and a small sclerotized head partly withdrawn into the broadly transverse thorax. Legs are greatly reduced and sometimes atrophied. The extreme type of larval modification is in weevils, where life occurs amidst immediately available food. Usually crescentic in form and entirely apodous, these larvae are mostly eyeless, with the antennae reduced to small papillae and there are no cerci.

Coleoptera are divided into three suborders, the Adephaga, the Polyphaga, and the very small, but archaic group, the Archostemata which somewhat resemble the first in the adult and the second in the larva. The **Adephaga** usually have filiform antennae and the hind coxal cavities are so large that they completely divide the first abdominal sternum: the hind wing has a short rectangular cell, the oblongum. The larvae (Fig. 80 A) are campodeiform with separate tarsi and usually paired claws. Included in this suborder is the single superfamily Caraboidea, whose members are mainly predators both as adults and larvae. Most of the terrestrial forms belong to the large family Carabidae (Ground Beetles), including the Cicindelinae (Tiger Beetles). Many aquatic Caraboids are included in the Dytiscidae while the Gyrinidae (Whirligig Beetles) are much modified for skimming on the surface of water. The **Polyphaga** include the greater part of the order. The antennae are varied in character: often clubbed, lamellate, or geniculate. The hind coxal cavities do not completely divide the first apparent abdominal sternite and the hind wings lack an oblongum and have few cross-veins. The larvae are of very varied types but have a combined tibiotarsus with a single claw. The Hydrophiloidea, with one family, are mostly aquatic; the maxillary palps are lengthened and partly replace in function the antennae which handle bubbles of air for respiration. The Staphylinoidea includes the Staphylinidae (Rove Beetles) and the Silphidae (Carrion Beetles) which have the elytra more or less shortened and the abdomen exposed. The superfamily Scarabaeoidea is an easily recognized group since its members have fossorial fore legs (Fig. 79 E) and a lamellate antennal club of platelike components (Fig. 79 F): their larvae are of the scarabaeiform type (Fig. 80 D). Included here are the Lucanidae or Stag Beetles and the Scarabaeidae or Chafers and Dung Beetles. The Buprestoidea includes one family of metallic, often green or blue, creatures; their larvae are legless borers living beneath the bark of trees and notable for their greatly widened prothorax. In the somewhat similar Elateroidea, the largest family is the Elateridae which include the tropical 'fire-flies' and the more numerous Click Beetles whose larvae or 'wire-worms' (Fig. 80 E) are destructive root-feeders of crops. The Dermestoidea includes the family Dermestidae (Larder Beetles) with densely hairy larvae which often damage wool and other dry animal materials. The very large group, the Cucujoidea, includes a clavicorn series of families, many of which have a marked antennal club. Typical of these are the Coccinellidae (Ladybirds) that are mostly beneficial since their larvae (Fig. 80 B) and adults prey largely upon aphids. The heteromerous series of families includes a great variety of forms whose main common feature is that the hind tarsi have 4 segments while the others have 5. Some of the most important families are the Tenebrionidae, which includes the Mealworms (*Tenebrio*) and Flour Beetles (*Tribolium*), and the Meloidae. To the last-named family belong the Blister Beetles, whose blood usually contains a caustic or blistering agent termed cantharidin. This product is chiefly prepared from the dried elytra of the 'Spanish Fly', *Lytta* (*Cantharis*) *vesicatoria*. *Meloe* and its allies undergo hypermetamorphosis (p. 128) and their larvae are parasitic within the nests of solitary bees or in the egg-masses of grasshoppers; the adults are known as Oil Beetles. The extensive superfamily Chrysomeloidea is characterized by the third tarsal segment being either bilobed or dorsally grooved and receiving the minute fourth segment at its base. The largest family is the Chrysomelidae or Leaf Beetles with nearly 30 000 species. Some, such as the Asparagus Beetle (*Crioceris asparagi*), the Flea Beetles (*Phyllotreta*, etc.) and the Colorado Beetle (*Leptinotarsa decemlineata*), are destructive to crops. The last-named was accidentally introduced into France in 1922 and has since greatly extended its range on potatoes in western Europe. The Cerambycidae or 'longicorns' comprise most of the other Chrysomeloids: they are forest insects whose larvae tunnel into the wood of trees. In the superfamily Curculionoidea or Weevils the head is usually produced into a well-marked rostrum (Fig. 79 D) and the tarsi are apparently 4-segmented. Their most important family is the Curculionidae with about 35 000 known species. The rostrum enables the female to bore holes in the medium in which the eggs are deposited. Included here are many injurious kinds such as the Grain Weevils (*Sitophilus*), the Cotton Boll Weevil (*Anthonomus grandis*), the Pine Weevil (*Hylobius abietis*) and others. The closely related Scolytinae (often treated as a distinct family) or Bark Beetles have a very short broad rostrum and include many species which rank as major forest pests.

ORDER 23. **STREPSIPTERA** (*strepsis*, a twisting; *pteron*, a wing)

Minute insects: males with branched antennae and degenerate biting mouthparts: fore wings modified into small club-like processes: hind wings very large, plicately folded. Females almost always degenerate parasites within the bodies of other insects. Stylops.

In their larval stages Strepsiptera are endoparasites mainly of certain aculeate Hymenoptera and Hemiptera-Homoptera. The first instar larvae are active creatures termed triungulins; on meeting a host they bore within and undergo hypermetamorphosis (p. 128). The females remain permanently as parasites with the fused head and thorax protruding between adjacent abdominal segments of the hosts. In a very few cases the females are active and free living. The males are short-lived and they fly on to the host to mate with the females. Parasitized hosts are said to be 'stylopized', the term being derived from the generic name *Stylops*. Bees of the genus *Andrena* are often affected and the parasitization tends to cause the hosts concerned to acquire altered sexual and other characters. Similarities in the structure of the male and the larva to the Coleoptera suggest the affinities of the order, sometimes indeed regarded as merely a superfamily of beetles. The resemblance in habits to the Meloidae, etc., is probably due to convergence.

ORDER 24. **MECOPTERA** (*mekos*, length; *pteron*, a wing)

Soft-bodied insects with two pairs of elongate similar wings: venation primitive, costal veinlets few. Head prolonged into a beak: mouthparts for biting: short cerci present. Larvae oligopod or polypod: pupae in earthen cells. Scorpion-flies.

A small order of about 300 species but with a world-wide range. Its members are usually recognizable by the beak-like head, maculated wings and the prominent external genitalia of the male. They are separable from Neuroptera by the small number of costal veinlets, the dichotomously branched Rs and by the undivided Cu_1. The habit of the males carrying the end of the abdomen upturned has given the name of Scorpion-flies to species of *Panorpa* and their allies. The imagines prefer shaded places and both they and the larvae are mainly carnivorous. The larvae live in litter, etc.: those of *Panorpa* bear three pairs of thoracic and eight pairs of abdominal limbs, while in *Boreus* only thoracic legs are present. There are only four British species which belong to *Panorpa* and *Boreus*. The last named has vestigial wings and its larva is exceptional in feeding upon mosses. Six genera and about 50 species occur in North America. Numerous fossilized wings from Permian rocks are referred to the Mecoptera, and the order is claimed to be closely related to forms that are ancestral to most of the Endopterygota.

ORDER 25. **SIPHONAPTERA** (*siphon*, a tube; *apteros*, wingless)

Very small, apterous, laterally compressed insects whose adults are ectoparasites of warm-blooded animals. Mouthparts for piercing and sucking. Larvae vermiform: pupae exarate in silken cocoons. Fleas.

The Siphonaptera are readily separated from other apterous parasitic insects owing to their being laterally, and not dorso-ventrally, flattened. Ocelli are present or absent and the antennae are short, 3-segmented organs reposing in grooves. The mouthparts (Fig. 81) are for piercing and have features in common with those of the lower blood-sucking Diptera. The stylet-like laciniae are finely denticulate and are the actual piercing organs. They are closely apposed to the slender epipharynx to form the food channel. Ventrally the edge of each lacinia is grooved and the apposition of the two grooves forms a salivary canal which receives a small process of the hypopharynx bearing the end of the salivary duct. The stipites are short triangular blades and the palpi are 4-segmented. The labium is a small basal plate bearing elongated palpi composed of a variable number of segments: their concave inner surfaces enable them to ensheath the laciniae. Fleas are covered with a tough cuticle and the legs are adapted for clinging and leaping. The maximum vertical leap of the human flea (*Pulex*

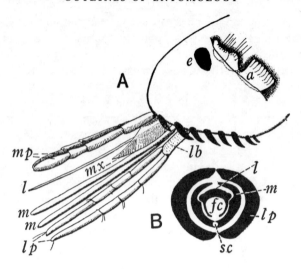

Fig. 81. Siphonaptera

A, head of flea showing mouthparts. B, transverse section of mouthparts. *a*, antenna; *f.c*, food canal; *l*, epipharynx; *lb*, labium; *lp*, labial palp; *m*, lacinia; *mx*, stipes; *mp*, maxillary palp; *sc*, salivary canal; *e*, ocellus

irritans) is stated to be $7\frac{3}{4}$ inches. All fleas are blood-sucking ectoparasites of birds and mammals and rarely exceed 4 mm in length. Usually each species has its particular host, but many can live at least temporarily on some other host. Thus the rat-flea, *Xenopsylla cheopis* (Fig. 82), frequently migrates to man and is the most potent vector in the transmission of the bacillus of bubonic plague which affects both rodents and man. The eggs of fleas are normally found in the haunts or sleeping places of the hosts. The larvae are whitish and vermiform, with a well-developed head bearing biting mouthparts and 13 trunk segments: they feed on particles of organic matter found in the hosts' lair or, in the case of the human flea, upon such matter among the dust and dirt of floors. The pupae are exarate and are enclosed in silken cocoons. About 1400 species of fleas are known.

The order is an isolated one, but it shows some affinity with Diptera. This is borne out in (1) the nature of the mouthparts; (2) the number of Malpighian tubes being four, as in most Diptera; and (3) the larvae resembling those of Nematocera.

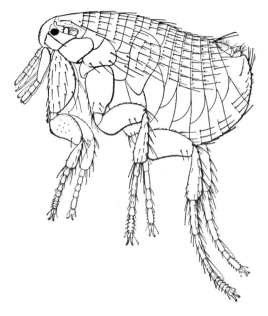

Fig. 82. *Xenopsylla cheopis*, male, × 20
(*After* Waterston. Reproduced by permission of the Trustees of the British Museum)

ORDER 26. **DIPTERA** (*dis*, two; *pteron*, a wing)

Moderately sized to very small insects with a single pair of membranous wings: hind pair modified into halteres. Mouthparts for sucking or for piercing also and usually forming a proboscis. Larvae vermiform: terrestrial, aquatic or parasitic: pupae weakly obtect or exarate and in a puparium: usually no cocoon. Flies.

Most flies are diurnal and many visit flowers for nectar, while numerous others feed upon decaying organic matter and diverse fluid substances. There is also a number of flies that are predators on smaller insects or have acquired blood-sucking habits. The single pair of wings is borne on the mesothorax and the metathoracic wings are modified into *halteres*, or balancers (p. 36). The mouthparts differ greatly in various families but in most cases the elongate labium forms the chief part of the proboscis.

A prevalent type of mouthparts is seen in the Blow-fly (*Calliphora*). Here the proboscis (Fig. 83) consists of a broad, basal *rostrum*, shaped like an inverted cone, and a distal *haustellum* bearing a pair of oral lobes or *labella*. The *rostrum* is formed by the clypeal region combined with the basal parts of the maxillae and labium. Situated in the rostrum is a cuticular stirrup-shaped framework or *fulcrum*. On the anterior surface of the rostrum the fulcrum presents an inverted V-shaped sclerite (Fig. 83 c), which is probably a derivative of the original *clypeus*. The foot piece of the stirrup strengthens the posterior wall of the cibarium and lateral extensions unite it with the clypeus. The clypeus and adjacent parts give origin to the dilator muscles of the cibarium and to the flexor muscle of the labrum. In front of the fulcrum are the *maxillary palpi*, and on either side is a darkly sclerotized rod-like *apodeme* which articulates with the labrum. The cibarium is closely attached to the posterior wall of the fulcrum and just in front of the latter it becomes enclosed in a small U-shaped theca which keeps its cavity distended. The *haustellum* probably represents the prementum and distal parts of the mentum. Its anterior surface is inflected to form a median *labial groove* or gutter. Posteriorly the haustellum is strengthened by a large concave plate, the *prementum*. The labial groove is sclerotized and each side is supported by a rod-like *paraphysis* (Fig. 83 A and B). The groove is roofed in anteriorly, to a large extent, by the *labrum*, whose inner lining is a channel forming a half tube that is closed by the *hypopharynx*. In front of the labrum the labial groove is closed by folds of the integument and is continued to the *prestomum*.[1] This is an aperture that is bounded and kept open by the arms of the *discal sclerite*. Attached to the latter are the *prestomal teeth*. Proximally the discal sclerite is connected to the paraphyses of the labial groove. The main skeletal support of the labella are the two arms of the *furca*, whose base articulates with paired processes of the prementum. The membrane covering the distal surface of the labella contains a series of food channels or *pseudo-tracheae*. These channels are kept open by a framework of incomplete cuticular rings which give them the appearance of tracheae. Each ring is bifurcated at one end and flattened at the other, the flattened and bifurcated extremities alternating (Fig. 84

---

[1] Some authors regard the prestomum as the cleft between the labella and term the opening enclosed by the discal sclerite as the 'oral aperture'. This latter term is confusing since the functional mouth is the opening into the pharynx at the base of the fulcrum.

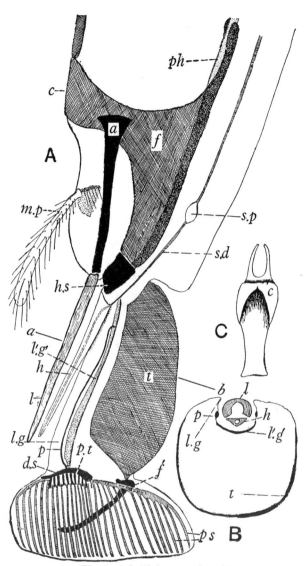

Fig. 83. *Calliphora* proboscis

A, lateral view, muscles omitted. B, transverse section of haustellum across line *a–b* in A. C, fulcrum, viewed from outer or anterior surface (drawn on smaller scale). *a*, apophysis; *c*, clypeus; *d.s*, discal sclerite; *f*, fulcrum; *f'*, furca; *h*, hypopharynx; *h.s*, theca; *l*, labrum; *l.g*, side wall of labial groove; *l'.g'*, sclerotized part of labial groove; *m.p*, maxillary palp; *p*, paraphysis (left); *ps*, pseudotracheae; *p.t*, prestomal teeth; *s.d*, salivary duct; *s.p*, salivary pump; *t*, prementum

D–F). The pseudotracheae open on the external surface of the labella through the cleft at the forked extremities of the cuticular rings. The pseudotracheae all converge to the prestomum and there are three sets of these channels to each labellum, i.e., those that run into an anterior collecting canal, those that run into a posterior collecting canal and a group between the two whose components open directly between the prestomal teeth which serve as conducting channels. When the proboscis is protruded the rostrum is extended through the expansion of the lateral air-sacs at

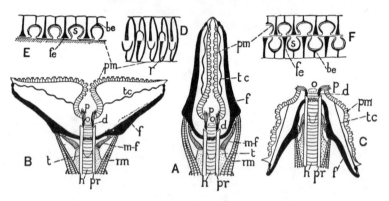

Fig. 84. Diagrams of the positions assumed by the labella of *Calliphora* A, resting position. B, filtering position. C, direct-feeding position. D, inner or dorsal view of portion of a pseudotrachea showing 5 rings. E, lateral view showing pseudotracheal membrane applied to a feeding surface. F, outer or ventral view of portion of a pseudotrachea. *be*, bifid end of ring; *d*, discal sclerite; *f*, arm of furca; *f.e*, flattened end of ring; *h*, sclerotized floor of labial groove; *m–f*, bar connecting prementum *t* with *f*; *o*, prestomum ('oral aperture' of authors); *p.t*, prestomal teeth; *pr*, paraphysis; *p.m*, pseudotracheal membrane; *r.m*, retractor muscle of furca; *s*, interbifid space; *t*, prementum; *t.c*, tendinous cord. (*Adapted from* Graham-Smith)

its base and probably of certain of the cephalic air-sacs also. The haustellum is brought into use by the contraction of its extensor muscles, and finally the labella are extended and rendered turgid by blood pressure. The retraction of the proboscis is mainly effected by the contraction of its numerous muscles. In the *resting position* the labella are flaccid and the two labella are in apposition (Fig. 84 A). When the fly is feeding on fluids the labella assume a phase termed by Graham-Smith the *filtering position* (B) and are pulled apart by the retractor muscles of the furca. The simultaneous injection of blood into the cavity of the labella converts the pseudotracheal surface into a pad which accommodates itself to unevenness of any surface to which it is applied. By means of the pumping action of the cibarial muscles suction is set up and liquid food is filtered in through the pseudotracheae to the prestomum. From here it passes along the channel formed by labrum and hypopharynx and so into the alimentary canal. In the *direct feeding position* (C) the lateral arms of the furca and the labella are pulled upward against the sides of the haustellum, thus completely exposing the prestomum. In a slightly less reflected condition the prestomal teeth project vertically downward, but the prestomum is not exposed. This phase is the *scraping position* of Graham-Smith, which enables the insect to rasp particles of sugar and other substances. Whereas in position B only fluids, and particles of a diameter not exceeding 0·006 mm, can enter the pseudotracheae, in position C particles of much larger size can be taken in directly along with fluids.

The mouthparts of *Calliphora* are highly specialized, and to determine the homologies of their components it is necessary to examine those of the lower Diptera. Mandibles are only present in blood-sucking Nematocera and Brachycera and are usually confined to the female. The maxillae in these suborders exhibit almost all the usual parts and comparisons indicate that the apodemes of *Calliphora* are probably derivatives of the stipes. The labella are regarded as modified labial palpi and, in this connexion, it may be noted that in some forms they are 2-segmented. In mosquitoes (*Anopheles*, *Culex*, etc.) the mouthparts in the female are very slender organs lodged in the labial groove (Fig. 85). The mandibles and the laciniae of the maxillae are modified into piercing stylets; the maxillae, being the stronger organs, are used for piercing. During feeding the labium becomes looped backward to allow of the maxillae perforating the skin of the host. The labrum is then inserted into the puncture and with the labium forms the food canal. The hypopharynx is concerned with the ejection of saliva which apparently acts as an anticoagulin on the blood. The part played by the mandibles seems to be of secondary importance. In the males both mandibles and maxillae are much reduced in *Anopheles* and are even more so in *Culex*: in feeding habits they are non-piercing, the apex of the labrum being merely dipped below the surface of liquid food so that the nutriment can be imbibed through the tubular labrum. Some of the Cyclorrhapha

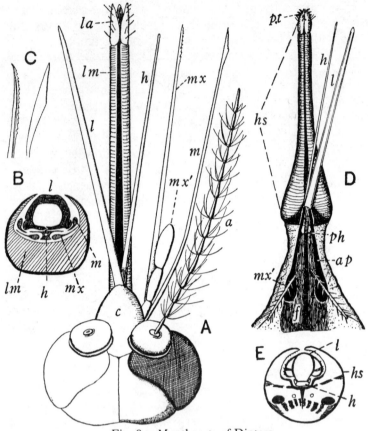

Fig. 85. Mouthparts of Diptera

A, mouthparts of *Culex* (female). B, transverse section of same. C, apices of maxilla and mandible respectively. D, *Stomoxys* mouthparts. E, transverse section of same. *a*, antenna; *ap*, apophysis; *c*, clypeus; *f*, fulcrum; *h*, hypopharynx; *hs*, haustellum; *l*, labrum; labellum; *lm*, labium; *m*, mandible; *mx*, maxilla; *mx′*, maxillary palp; *ph*, cibarium; *p.t*, prestomal teeth

have also acquired blood-sucking habits which prevail in individuals of both sexes, notably in the Biting Housefly (*Stomoxys*), the Tsetse flies (*Glossina*), the Forest-fly (*Hippobosca*), etc. The haustellum (Fig. 85) in such cases is a rigid, horny, piercing organ devoid of pseudotracheae and there are no mandibular or maxillary stylets. The puncture is made by the strong prestomal teeth and the blood is drawn in through the prestomum into the food canal formed by the combined labrum and hypopharynx.

The Horse-flies (Tabanidae) combine the filter feeding method of the blow-fly with the piercing method of mosquitoes. Pseudo-tracheae are present on the labella and, in the females, the mandibles and laciniae are broad stylets used for piercing in order to obtain blood.

The *ptilinum* or frontal sac is a cephalic organ found in the most highly specialized flies. Its presence is shown externally by the U-shaped *ptilinal suture* embracing the insertions of the antennae (Fig. 86 I). The suture is an extremely narrow slit along the margins of which the integument is invaginated as a sac or ptilinum. The latter is everted through the suture prior to the emergence of the fly. With the aid of the ptilinum the insect ruptures the puparium. The ptilinum is everted by blood pressure and muscular action and, having served its purpose, it becomes withdrawn into the head cavity, where it remains (Fig. 86 J). It can, however, be everted again by squeezing the thorax of the young fly with forceps so as to force the blood forward.

The thorax of Diptera is characterized by the great size of the middle segment bearing the wings and the correlated reduction of the segments in front and behind. The most primitive venation occurs in crane-flies and some other Nematocera, the veins being predominantly longitudinal with but few cross-veins (Fig. 86 B). The narrow wing-bases have led to a great reduction of the anal veins and Cu_2 is absent or vestigial. Specialization is by reduction and is most pronounced in Cyclorrhapha (Fig. 86 A). An ovipositor is rarely developed, but in *Musca*, and many other flies, the end segments of the abdomen form a telescopic tube serving the same purpose.

Dipterous larvae never have thoracic legs and are typically amphipneustic or often metapneustic: propneustic or apneustic forms occur more rarely. Three thoracic and nine abdominal segments are present, and in Nematocera the head is fully developed (*eucephalous*). The mouthparts are less modified than in other groups and the paired components work horizontally. In the Brachycera the head is incomplete posteriorly and partly em-bedded in the prothorax (*hemicephalous*): the mouthparts are highly modified and work in the vertical plane. Among Cyclor-rhapha (Fig. 86) the larvae are *acephalous* and this condition results from the whole head being invaginated into the thorax. During this process the mouth becomes carried far inward and communication

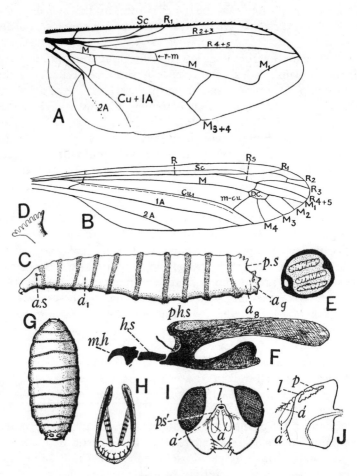

Fig. 86. Diptera

A, *Musca domestica*, right wing. B, *Tipula*, right wing. C–G, *Calliphora*. C, larva.
D, anterior spiracle and, E, posterior spiracle of same. F, cephalopharyngeal
skeleton. G, puparium. H, transverse section of pharynx of larva of *Sarcophaga*.
I, frontal view of head of fly of suborder Cyclorrhapha. J, lateral and sectional view
of the same. a_1, a_8, a_9, abdominal segments; a, antenna; $\acute{a}$, arista; *a.s,* anterior
spiracle; *d.c,* discal cell; *h.s,* hypostomal sclerite; *l,* lunule; *m.h,* mouth hook; *p,*
ptilinum (retracted); *p.s,* posterior spiracle; *p.s′,* ptilinal suture; *ph.s,* pharyngeal
sclerite

with the exterior is by means of a secondary passage or *atrium*. The apparent head rudiment is in reality a circular papilla-like fold of the neck. True mouthparts are atrophied and their place is taken by adaptive structures that form the *cephalopharyngeal skeleton* (Fig. 86 F). Of these the paired *mouth hooks* are alone freely movable and work in the vertical plane: they articulate with a *hypostomal sclerite* whose two components are joined by a crossbar. This sclerite articulates in turn with the *pharyngeal sclerite* that is composed of two vertical lamellae united below to form a trough for the support of the pharynx (Fig. 86 H). The pharyngeal sclerite is the homologue of the fulcrum of the imago. The mode of life of Cyclorrhaphous larvae can be ascertained from the mouthparts. In carnivorous forms the mouthparts are sharply hooked, whereas in phytophagous forms they are toothed. Also, the pharyngeal floor is ridged in saprophagous forms, less so or smooth in phytophagous forms and wholly without ridges in carnivores.

Dipterous larvae pass through three or four instars and the pupae are exarate or weakly obtect. In the Cyclorrhapha the cuticle of the third instar is not shed, but hardens to form a shell or *puparium* enclosing the pupa (Fig. 86 G). The puparium ruptures along special lines of fracture owing to pressure exerted from within. In the higher Cyclorrhapha the inflated ptilinum forces open the puparium and so liberates the imago.

About 64 000 species of flies are known, and of these more than 5200 kinds inhabit the British Isles. The order is one of very great economic importance either as larvae or as adults. The pathogenic organisms of some of the most virulent diseases such as malaria, sleeping sickness, elephantiasis and yellow fever are transmitted to man through the agency of blood-sucking flies. The housefly and its allies act as mechanical carriers of disease germs, and in this way contaminate human foods. Various Dipterous larvae induce diseased conditions in the bodies of man and domestic animals, and such infections are included under the term *myiasis*. Larvae of other species are injurious to the agriculturist and their activities result in great financial losses. The depredations of members of this order are offset, to a considerable degree, by those species of carnivorous habit that destroy large numbers of noxious insects. Some of these are predators either as larvae or adults, while many others are endoparasites in their larval stages.

Diptera are classified into three suborders, viz.: **Nematocera,**

Brachycera and **Cyclorrhapha.** The **Nematocera** have many-jointed antennae, usually longer than the head and thorax, and the maxillary palpi have 4 or 5 segments. The larvae are usually eucephalous with horizontally biting mandibles. The slender, long-legged Craneflies or Tipulidae are familiar creatures. The larvae are metapneustic and those of some species of *Tipula*, known as 'leather-jackets', are injurious to the roots of pasture grasses and crops. The Chironomidae or Midges have mostly aquatic larvae that are apneustic. In some species the larvae are known as blood-worms owing to the presence of haemoglobin in the blood plasma. The Culicidae or Mosquitoes are very slender insects with long piercing mouthparts and the wing margins, together with the veins, are clothed with scales. All the immature stages are aquatic (p. 131), the larvae being metapneustic. With few exceptions, female mosquitoes are able to pierce the skin of vertebrates and feed upon the blood; they also feed upon various plant juices and some kinds may never taste blood at all. The larval habits are varied: some inhabit shady pools, others are found in streams, dykes, tree holes, salt marshes, etc. Some 32 species of mosquitoes occur in Britain, and of these the commonest is *Culex pipiens*. Among other families of Nematocera are the Simuliidae or Buffalo Gnats, the Mycetophilidae or Fungus Gnats, and the Cecidomyidae or Gall Midges. Included in **Brachycera** are 14 families of stout-bodied flies with short antennae, generally 3-segmented and often with the last segment prolonged into a style. The maxillary palpi are 1- or 2-segmented and the larvae are hemicephalous with vertically biting mandibles. Mention may be made of the Tabanidae or Horse-flies, the females of which are blood-sucking in habit, and the Asilidae or Robber-flies, which are predators on other insects.

The **Cyclorrhapha** comprise all the higher Diptera. Their antennae are 3-segmented with a dorsal bristle-like *arista* (Fig. 86 I) and the maxillary palpi are 1-segmented. The larvae are acephalous and commonly amphipneustic; their mandibles are replaced by mouth hooks movable only in the vertical plane and the pupae are enclosed in a puparium. Among this great assemblage of forms the Syrphidae or Hover-flies are without a ptilinum and their larvae include many predators of aphids. Most families possess a ptilinum, but only a few can receive separate mention. The Tephritidae or Fruit-flies include destructive larvae that mine the pulp of economic and other fruits. The Drosophilidae or Pomace-flies have saprophagous larvae and the Oestridae, which include the Warble-flies and Bot-flies, have larvae that are endoparasites of mammals. The Muscidae include the Housefly (*Musca domestica*) and its allies, together with blood-sucking forms such as the Stable-flies (*Stomoxys*); allied are the Tsetse-flies (*Glossina*). The Tachinidae are very bristly flies whose larvae are endoparasites of other insects. The Calliphoridae are either similar parasites or, like *Calliphora, Lucilia,* etc., have saprophagous larvae. The Hippoboscidae are viviparous and live, as adults, as blood-sucking ectoparasites on birds and mammals: the wingless Sheep Ked (*Melophagus*) and the Forest-fly (*Hippobosca*) are well-known examples.

ORDER 27. **LEPIDOPTERA** (*lepis,* gen. *lepidos,* a scale; *pteron,* a wing)

Small to very large insects clothed with scales. Mouthparts with galeae usually modified into a spirally coiled suctorial proboscis: mandibles rarely present. Larvae phytophagous, polypodous: pupae obtect or partially free, usually in cocoons. Butterflies and Moths.

An immense order with over 140 000 species which have the wings, and usually the body and appendages, more or less covered with pigmented scales (pp. 17, 29): over the wing surfaces the scales give rise to characteristic colour patterns. The structural similarity of these insects has led to great uniformity of behaviour. The imagines live upon nectar, over-ripe fruit, honeydew, etc., while their larvae with few exceptions feed entirely upon phanerogamic plants – leaves, roots, seeds, wood. In *Micropterix* and its allies, which are pollen feeders, functional and complete mandibles and maxillae (Fig. 87 E) are present. In *Eriocrania* mandibles are reduced, the maxillae have lost the laciniae, while the galea of either side is grooved and functions with its fellow as a suctorial proboscis; in the pupa, however, there are functional mandibles. In the rest of the Lepidoptera mandibles are vestigial or wanting and the proboscis (Fig. 87 D) may attain a length greater than that of the insect. Each galea is a tube whose cavity is continuous with that of the head: its flexibility is caused by a series of rings separated by membrane. The two components of the proboscis interlock and thus enclose a median food canal (Fig. 87 C). The actual intake of food is effected by the action of a buccopharyngeal pump.

Extension of the proboscis is caused by the contraction of three pairs of extensor muscles which exert pressure on the blood by reducing the cranial cavity. This results in blood being forced through a valve in the cavity of the stipes and so into the interior of each half of the proboscis, thus extending the organ as a whole. As R. E. Snodgrass has stated, the extension of the proboscis by blood pressure has its analogy in the unrolling of a toy paper 'snake' by inflating it. Relaxation of the extensor muscles results in a backward flow of blood from the proboscis and the coiling of the latter is brought about by the contraction of the numerous oblique muscles that cross the cavity of each component (Fig. 87 A, B). A reduced labrum, bearing lateral lobes or pilifers, overlies the base of the proboscis. Except in some primitive families maxillary palpi are much reduced or absent. The labium is rep-

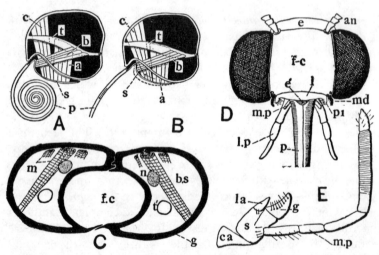

Fig. 87. Lepidoptera

A, diagram of action of proboscis: right half of head in section with proboscis retracted and coiled. B, the same with proboscis extended by contraction of extensor muscles, a, b and c, which compress the stipes and force blood through a valve into the proboscis. C, transverse section of proboscis. D, frontal view of head and mouthparts of a moth. E, right maxilla of *Micropteryx*. (A and B *adapted from* J. B. Schmitt.) *an*, base of antenna; *b.s*, blood space of proboscis; *ca*, cardo; *e*, epicranium; *e'*, epipharynx; *f.c*, food channel; *f–c*, frontoclypeus; *g*, galea; *l*, labrum; *la*, lacinia; *m*, retractor muscle; *md*, mandible (vestigial); *mp*, maxillary palp; *n*, nerve; *p*, proboscis; *pi*, pilifer; *t*, tentorium; *t'*, trachea

resented by a simple plate bearing 3-segmented palpi that project conspicuously on either side. Numerous moths take no food and their mouthparts consequently display varying degrees of atrophy. The prothorax bears a pair of erectile lobes or *patagia*, well displayed in many Noctuidae, etc.; tegulae (p. 29) are well developed and characteristic. The venation (Fig. 88) of the most primitive families is of a generalized type with very few cross-veins. Specialization involves the ultimate disappearance of Cu_2 from both pairs of wings, the reduction of Rs to a single branch in the hind wing and the formation of a large *discal cell* in each wing by the absorption of intervening veins and cells. The prevalent wing-coupling apparatus shows sexual dimorphism. In the male the *frenulum* (p. 30) is single and a hook-like *retinaculum* is usually present beneath the base of Sc (Fig. 88 B); in the female the frenulum is commonly formed of several bristles and the retinaculum is on Cu_1. The cubital retinaculum may also be present in the male. In the Hepialidae (Fig. 88 A) and some other primitive moths a process or *jugum* arises from the base of the fore wing: in flight, the jugum lies on top of the base of the hind wing. In butterflies and certain moths, that have lost the frenulum, *amplexiform coupling* obtains: the enlarged humeral lobe of the hind wing is maintained against the stiffened base of the fore wing, thus ensuring synchronous action of the two wings. It may be added that wings are vestigial or absent in the females of certain families, notably the Psychidae and a few Geometridae. On either side of the metathorax, or the base of the abdomen, in many moths there is a complex *tympanum* (p. 60), its presence or absence being constant for large groups of families.

The larvae or caterpillars have a well-developed head, 3 thoracic and 10 abdominal segments (Fig. 54 B). Spiracles are present on the prothoracic and first 8 abdominal segments. The mouthparts are masticatory but reduced; the antennae are small 3-segmented organs, and behind them is a group of 6 ocelli on either side. Each thoracic segment bears single-clawed legs and a pair of abdominal feet is present on segments 3 to 6 and 10. These organs are fleshy projections whose grasping surface is armed with hooks or crochets that are arranged in circles in the lower families but restricted to an arc or band in the more specialized groups.

In the Geometridae abdominal feet are present only on segments 6 and 10, their caterpillars being known as 'loopers' from their

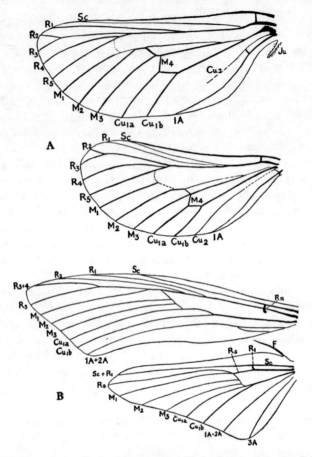

Fig. 88. Venation of, A, Monotrysia (Hepialidae) and, B, Ditrysia (Sphingidae)

Ju jugum; F, frenulum; Rn, retinaculum

method of crawling. Most caterpillars are protected either by their cryptic form and colour, or by the display of warning patterns, or by the adoption of concealed habits. The labial glands (p. 97) are modified into silk glands whose secretion is emitted through a median spinneret associated with the hypopharynx. Silk glands may be several times longer than the body in species that form dense cocoons: in all cases the salivary function is performed by mandibular glands.

The pupa in the lower Lepidoptera Ditrysia has the appendages free and most of the abdominal segments movable: aided by a spiny armature, such a pupa usually issues partially from the cocoon to allow of the emergence of the imago. In the higher forms the pupa is obtect (p. 127) with only three free abdominal segments and remains attached to the cocoon by means of a terminal hooking device or *cremaster*. Many butterflies have naked and protectively coloured pupae with the cocoon reduced to a pad of silk to which the cremaster is hooked.

Economically the order is of great importance owing to the damage incurred by the feeding activities of the caterpillars. Only a few examples will be quoted. Species of *Pieris* or the 'White' butterflies are major pests of cruciferous and other crops. The Gipsy Moth (*Lymantria dispar*) and the Nun Moth (*L. monacha*) are great defoliators of forest trees: the European Corn Borer (*Ostrinia nubilalis*) is destructive to maize, etc., in North America: the Codling Moth (*Cydia pomonella*) is a widespread enemy of the apple, and the Mediterranean Flour Moth (*Ephestia kuehniella*) is an almost universally distributed pest in flour mills, etc. Two other pests of very wide range are the Pink Bollworm (*Platyedra gossypiella*) of the cotton plant and the Angoumois Grain Moth (*Sitotroga cerealella*) which infests wheat, maize, etc. Mention also needs to be made of the Clothes Moths (*Tinea pellionella*, *Tineola bisselliella* and *Trichophaga tapetzella*) which attack woollen clothing, rugs, furs, etc. To offset this, the Silk Moths *Bombyx mori* and certain Saturniidae are beneficial in that they provide commercial silks.

Lepidoptera are closely related to the Trichoptera, the two orders being derived from a common ancestor. The complete M_4 in the fore wing of archaic Trichoptera and the absence of broad scales, of biting mouthparts and of a cloaca in the female will distinguish them from the Micropterigidae. In larval structure and habits the two orders are widely divergent, the Trichoptera being essentially aquatic and the Lepidoptera terrestrial. The Lepidoptera are classified into three suborders. The Zeugloptera includes only the Micropterigidae, primitive in their venation and biting mouthparts but in the female with vagina and rectum opening into a common cloaca on the ninth sternite. The Monotrysia, have, at least in rudiment, the proboscis of higher Lepidoptera, but the female has either a cloaca or genital openings on the ninth sternite.

The venation is often primitive, as in the Eriocraniidae and Hepialidae, but may be reduced. The Ditrysia includes more than 98% of the order: their venation is notably different in the two pairs of wings, Rs being reduced to a single vein in the hind wings (Fig. 88 B). In the female, there is a copulatory aperture on sternite eight and separate anus and egg pore on sternite nine. A frenulum occurs in most species, but several groups have lost this structure and acquired the amplexiform type of wing-coupling. The Cossidae or Goat Moths have the most archaic venation. The Pyralidae include among their genera *Acentria* and *Nymphula*, whose larvae are exceptional in being aquatic. The Pterophoridae or Plume Moths have deeply fissured wings and the Tineidae are an extensive family of varied larval habits and include the Clothes Moths. The Saturniidae are notable for their dense silken cocoons, those of some species being of commercial value: *Attacus* includes the oriental Atlas Moths which attain a wing expanse of 25 cm. The Bombycidae include *Bombyx mori*, a native of China, whose larva is the well-known 'silk worm'. Butterflies form the superfamilies Papilionoidea and Hesperioidea, distinguished by the clubbed antennae and the absence of a frenulum. Other notable groups are the Geometridae, whose larvae are 'loopers', the Hawk Moths or Sphingidae and the great family of Owlet Moths or Noctuidae.

ORDER 28. **TRICHOPTERA** (*thrix*, gen. *trichos*, a hair; *pteron*, a wing)

Moth-like insects with two pairs of densely hair-covered wings showing predominantly longitudinal venation with few cross-veins. Mouthparts reduced: mandibles not functional. Larvae aquatic, generally in portable cases: thoracic legs and paired caudal appendages ending in hooks present. Pupae aquatic, with strong mandibles. Caddis-flies.

Trichoptera (Fig. 89) are weakly flying and mostly nocturnal insects, found usually in the vicinity of water. They are obscurely coloured, generally of some shades of brown or grey, and the wings are closed roof-like over the back when at rest. They are closely allied to the Lepidoptera (p. 195), the venation of the family Rhyacophilidae closely resembling that of *Micropterix* and allied moths. Caddis-flies only take liquid food, which is licked up by the

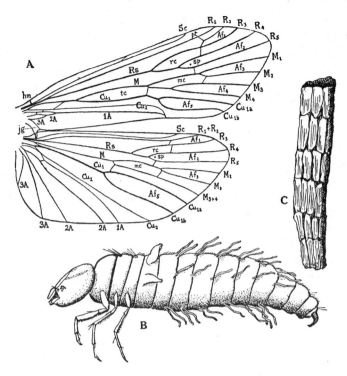

Fig. 89. Trichoptera

A, right wings of *Stenopsychodes* (Rhyacophilidae) (*after* Tillyard). B, *Phryganea*, larva, and C, its case

broad haustellum and traverses the channel formed by the labrum and the remainder of the hypopharynx. The maxillary lobes are galeae only, thus prefiguring the condition found in Lepidoptera. The eggs are laid in masses in or near water and are commonly protected by mucilage. The larvae all have the head well sclerotized with very small antennae and the first pair of legs shortest and stoutest; the abdomen is formed of nine segments, the last bearing a pair of jointed appendages ending in hooks. Caddis larvae present two main types: in the first type the head is inclined at an angle with the body and there are dorsal, lateral and ventral tufts of abdominal tracheal gills. Such larvae make portable cases of extraneous material bound together and lined with silk produced by the modified labial glands. These cases may be constructed of leaf- or stem-fragments, sand grains, empty shells of molluscs, etc., and are very constant in character for different genera. Dorsal and lateral papillae on the first abdominal segment maintain the larva in position within its case and allow an even flow of water through the latter. When walking, the head and the sclerotized first, or first and second, thoracic segments are protruded from the case, which is gripped by the caudal hooks and dragged along at the same time. In the second type the larvae are usually more active, with elongated body and the head prognathous. They seldom make cases and in many instances live in silken retreats. There are no papillae on the first abdominal segment and gills are usually wanting. The anal appendages are often well developed and are used to grip the silken tunnels or to hold on to rocks when they leave their retreats. Trichopterous pupae breathe cutaneously or by means of gills, as in the larva. They are protected either by the original but adapted larval cases, or by special shelters constructed for the purpose. Strong mandibles are present which enable them to cut their way out for the emergence of the imago. In many species the pharate adults are able to swim in order to reach the surface of the water. For this purpose they use the long middle legs of the pupa, which are fringed with swimming hairs. Fewer than 3000 species of Trichoptera have been described and, of these, over 180 species inhabit Britain.

ORDER 29. **HYMENOPTERA** (*hymen*, a membrane; *pteron*, a wing)

Minute to moderate-sized insects with membranous wings, the hind pair the smaller and connected with fore pair by hooklets: venation specialized by reduction. Mouthparts for biting and licking. Abdomen with 1st segment fused with thorax: a sawing or piercing ovipositor present, larvae usually polypod or apodous: pupae generally in cocoons. Sawflies, Ants, Bees, Wasps, Ichneumon flies and their allies.

The most constant distinctive feature of Hymenoptera is the fusion of the first abdominal segment or *propodeum* (Fig. 91 E) with the metathorax, which occurs in the prepupa but it is not very evident in the Symphyta. The propodeum, it will be noted, bears the first pair of abdominal spiracles. The mouthparts show the most generalized condition among sawflies. Mandibles are always present, while the maxillae and labium have all the usual com-

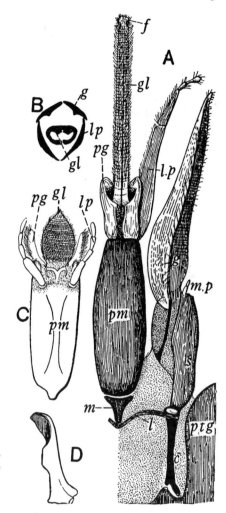

Fig. 90. Hymenoptera,
mouthparts

A, labium and maxilla of honey
bee (posterior or ventral view). B,
section across same in feeding
position. C, labium of a short-
tongued bee (*Sphecodes*). D, honey
bee, right mandible. *f*, flabellum;
g, galea; *gl*, glossa; *l*, submentum;
l.p, labial palp; *m*, mentum; *m.p*,
maxillary palp; *pg*, paraglossa; *pm*,
prementum; *ptg*, postgena; *s*,
stipes

ponents: the glossae are always fused and form a broad tongue.
Only small differences occur in the majority of adult Hymenoptera. In bees (Fig. 90), however, there is a progressive lengthening,
in different genera, of the glossa and associated parts to form, in
the higher types, a proboscis adapted to extract deeply-seated
nectar from flowers. In the honey bee the mandibles are smooth-
edged and used for manipulating wax and other purposes (Fig.

90 D). The maxillae are greatly elongated with rod-like cardines and the galeae are large thin blades much longer than the stipites; a pair of small membranous lobes probably represents the laciniae and the palpi are reduced to papillae. In the labium there is a long prementum articulating with a small mentum whose apex fits into the angle of a V-shaped suspensory sclerite or submentum (lorum) that articulates with the distal ends of the cardines. The glossa is greatly elongated and ends in a spoon-like lobe or *flabellum*: at the base of the glossa there are scale-like paraglossae. The labial palpi have the two basal segments flat and blade-like, leaving the distal segments unmodified. When the bee feeds on any easily accessible liquid the maxillary galeae and the labial palpi form an improvised tube along with the glossa (Fig. 90 B). The flabellum is immersed in the food, and by a rapid backward and forward motion of the glossa the liquid is drawn up the tube: it is then sucked up into the digestive canal by the action of the cibarial pump. Where the food is more inaccessible the glossa may be projected far beyond the ends of the maxillae. The ventral or posterior surface of the glossa bears a deep channel which reaches to the flabellum: the saliva traverses this channel and becomes mixed with the food during ingestion.

The venation (Fig. 91) deviates widely from the primitive type and during development the veins are demarcated before the tracheae develop. This fact, along with the frequent anastomosis of the veins, forming numerous cells, makes the homologies of the veins hard to ascertain. The most generalized condition occurs in the Symphyta and various stages in reduction prevail in other groups of the order. Bees and wasps show an intermediate condition of reduction while the extreme phases occur in the Parasitica where the veins may be restricted to the costal margin or be totally atrophied. Wing-coupling by means of costal hooklets of the hind wing engaging the reflected hind margin of the fore wing prevails throughout the order (Fig. 91 B).

An ovipositor is always present. Its lateral valves are represented by a pair of sensory palpi, the inner valves are fused to form a stylet and the anterior valves form a pair of lancets which have tongue and groove articulations with the stylet. In the Symphyta the ovipositor functions as a saw, both the lancets and the stylet being prominently toothed. Among the Apocrita, the ovipositor is a piercing organ in many of the Parasitica, while in the Aculeata

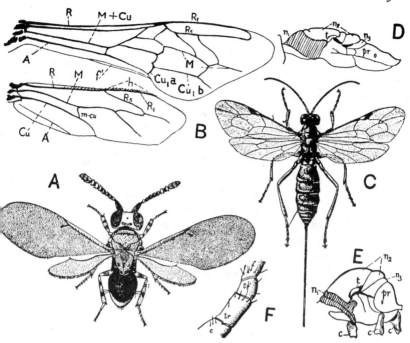

Fig. 91. Hymenoptera

A, a Chalcid (*Aphycus*). B, right wings of worker bee. C, an Ichneumon (*Exeristes*), female (*from Tech. Bull.* 460, *U.S. Dep. Agric.*). D, thorax of an ant, side view. E, do. of a worker bee. F, base of the leg of an Ichneumon. *c*, coxa; *f*, femur; *f'*, flange to hind margin of fore wing; *h*, hooks or hamuli; n_1–n_3, pro-, meso- and metanota; *pr*, propodeum; *t*, tegula; *tr*, *tr'*, trochanter and trochantellus

it is a sting (Fig. 11). In most Hymenoptera the eggs pass down the ovipositor channel and may become greatly compressed and stretched to allow of their free transit: in the stinging forms, on the other hand, the eggs are discharged from the genital pore at the base of the ovipositor and the latter organ acts as a poison-injecting instrument.

A Hymenopterous larva has a well-developed head, 3 thoracic and 9 or 10 abdominal segments. In the sawflies there are three pairs of thoracic legs and abdominal feet are present either on all the segments (*Xyela*) or, more usually, on segments 2 to 8 and 10. In other Hymenoptera the larvae are apodous (Fig. 54 E) but evanescent appendages may appear in the early instars of some Parasitica. There are usually 9 or 10 pairs of spiracles, except in endoparasitic larvae where the number is variable. A somewhat weak cocoon commonly encloses the pupa, but is wanting in the Chalcidoidea. Hymenoptera rank among the largest and most highly developed orders of insects and are of special interest from the wide range of biological features they display. In the great development of their instincts they stand in the forefront of all invertebrates, and their behaviour has been the subject of studies by some of the most famous naturalists. At least 100 000 species are known, and although the vast proportion of these are solitary in habit like other insects, individuals of some groups live together in great societies, as is the case with ants and certain bees and wasps (pp. 137–41). Hymenoptera are also remarkable for the highly evolved state parasitism has reached in the order: tens of thousands of species betray this habit and their larvae present special respiratory and other adaptations in accord with their modes of life. Associated with parasitism is the phenomenon of polyembryony (p. 108) which attains unique developments. Parthenogenesis is more frequent among Hymenoptera than in any other order of animals; besides being an important factor in social life, it may also be associated with alternations of generations.

From the economic standpoint the Hymenoptera confer many benefits upon man. Bees are important pollinators of fruit trees and other plants, while the honey bee is well known to yield honey and wax. The important part played by the parasitic Hymenoptera in destroying myriads of injurious insects is a recognized feature in the biological control of pests. Among noxious members of the order the majority are defoliating larvae of sawflies and the boring larvae of wood-wasps or Siricidae. Of lesser importance are the phytophagous larvae of certain Chalcids (p. 209).

Hymenoptera are divided into two suborders, Symphyta and Apocrita. In the **Symphyta** the abdomen is broad with no basal constriction or petiole behind the propodeum which is only partially amalgamated with the thorax (Fig. 92 B). The larvae are phytophagous and possess thoracic and usually abdominal feet (Fig. 92 C). Most of the species are included in the superfamily Tenthredinoidea, which comprises 6 families.

The Cephidae or Stem Sawflies are a small group whose larvae tunnel within the stems of graminaceous and other plants, while those of the Siricidae or Wood Wasps bore into the wood of trees. The true sawflies (Fig. 92 A) whose ovipositor acts as a

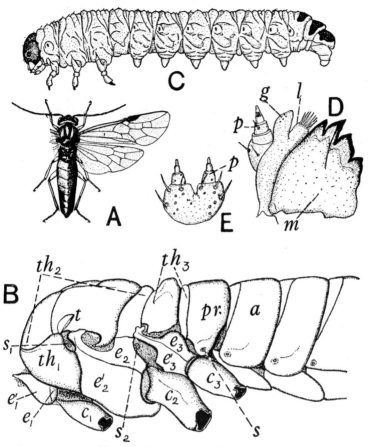

Fig. 92. Hymenoptera Symphyta

A, Larch Sawfly (*Pristiphora erichsonii*), female (*after Washburn*). B, side view of thorax and base of abdomen of a sawfly. C, larva of Plum Sawfly (*Hoplocampa flava*) in 4th instar, × 7. D, left mandible and maxilla of same, × 55. E, labium, × 55. (C, D and E *adapted from* H. W. Miles.) *a*, second segment of abdomen; *c*, coxa; *e'*, episternum; *e*, epimeron; *g*, galea; *l*, lacinia; *m*, mandible; *p*, palp; *pr*, propodeum; *s*, spiracle; *t*, tegula; th_1–th_2, thoracic segments. The numerals refer to the respective thoracic segments

saw in cutting shallow notches or deep incisions in plant tissues for placing their eggs mostly have leaf-feeding larvae. Their larvae are caterpillars which differ from those of Lepidoptera in the single ocellus on each side of the head, in the segmental positions of the abdominal feet and in the absence of crochets. The only other

Symphyta are *Orussus* and its allies which form the small super-family Orussoidea. Their specialized venation, slender retracted ovipositor and the unusual insertion of the antennae – below the clypeus and eyes – separate them from other Hymenoptera. Their larvae are legless ectoparasites of larvae of wood-boring beetles or Siricids.

The **Apocrita**, which comprise the majority of Hymenoptera, have the abdomen stalked or constricted between the propodeum and the true second segment. They are divided into two main groups, the Aculeata and the Parasitica, which, however, inter-grade both as regards structure and habit. In the **Aculeata**, or stinging forms, the eighth sternite is retracted, so that the ovi-positor appears to issue from the apex of the abdomen. They include a number of large superfamilies of which the Scolioidea are the least specialized. They make no real nest and have more or less parasitic habits. The ants or Formicidae are all social insects whose female loses her wings after mating and in most species is aided by numerous wingless workers. The periole behind the propodeum is raised into one or two nodes. The more primitive ants rear their larvae on insect prey: other ants store seeds, cultivate fungi or obtain honey-dew from aphids. The true wasps or Vespoidea include many solitary nest-making species nearly all of which store paralysed caterpillars for their offspring; one group, however, has social habits. The Sphecoidea or digger-wasps feed their larvae on paralysed insects of various orders or on spiders. The bees, or Apoidea, are closely related to them but many of their hairs are branched and the hind basitarsus is usually broad. Many bees make nests very like solitary wasps but they store a mixture of pollen and nectar for their larvae. A few groups of which the honey bees (or Apini) and the bumble bees (or Bombini) are the best known, are social.

In the **Parasitica** the trochanters are generally followed by a trochantellus (Fig. 91 F) and, in most cases, the ovipositor is exposed almost to its base, the eighth sternite not being retracted. They include an enormous number of species, small or minute in size, whose larvae are ecto- or endoparasites of other insects. The Ichneumonoidea include the largest forms and the venation is well developed with a pterostigma (Fig. 91 C); to a large extent the ichneumon flies and their allies parasitize caterpillars of Lepidop-tera. The remaining Parasitica are mostly small or minute insects with greatly reduced venation, generally with few or no closed cells. The Cynipoidea are best known from their gall-producing members, more than 80% of which are confined to species of *Quercus*. A considerable number of other Cynipoidea are endoparasites of Diptera. The Chalcidoidea include the largest number of species (Fig. 91 A). They are parasites and hyperparasites except for a small proportion that form galls or develop within seeds: the most notable of these is *Blastophaga*, associated with the pollination of the fig. The Proctotrupoidea include many minute egg parasites. They resemble the Aculeata in the concealed ovipositor. The parasitic habits of some of the Scolioidea connect the Aculeata with the Parasitica on biological grounds.

6

Relationships of Insects

INSECTS AND OTHER ARTHROPODS

The arthropods form the largest group in the animal kingdom and can be recognized by the following characters. The body is segmented and covered by a chitinous exoskeleton. A variable number of the segments carry paired, jointed appendages that exhibit functional modifications in different regions of the body. The heart is dorsal and is provided with paired ostia, a pericardium is present, and the body-cavity is a haemocoele. The central nervous system consists of a supra-oesophageal centre or brain connected with a ganglionated ventral nerve-cord. The muscles are composed almost entirely of striated fibres and there is a general absence of ciliated epithelium. The earliest known arthropods, the fossil group of Trilobita, were aquatic, but among living forms it is only the Crustacea that have retained this mode of life to a predominant extent. No other group of invertebrate animals has so large a proportion of its members terrestrial in habit.

Arthropods are divisible into six main groups as follows:

(a) Worm-like terrestrial forms bearing a pair of pre-antennae: the body unsegmented externally and with numerous pairs of unjointed legs. Onychophora.

(b) Bearing antennae, usually aquatic in habit and breathing either cutaneously or by means of branchiae. Trilobita, Crustacea.

(c) Bearing antennae, primarily terrestrial in habit and breathing by tracheae. Myriapoda, Insecta.

(d) Without antennae, usually terrestrial in habit and breathing by lung-books and tracheae. Chelicerata.

The phylogenetic relationships of these major groups have been much disputed, but there is now considerable evidence that arthropod characteristics arose independently in several evolutionary lines. One of these, the phylum now known as the Uniramia, includes the insects and their closest relatives, the Myriapods and the Onychophora, of which the latter represent the least modified survivors of the ancestral Uniramian stock. The Trilobita, Chelicerata and Crustacea, on the other hand, seem to be only rather distantly related to each other or to the Uniramia, and their origins are obscure.

The **Trilobita** were marine animals and their remains are numerous in Palaeozoic rocks of Cambrian and Silurian date. The body is highly specialized and divided longitudinally into median and lateral or pleural regions. The appendages, on the other hand, are very primitive and consist of a single pair of antennae almost certainly homologous with the crustacean first antennae. The remaining appendages are biramous and only slightly differentiated among themselves. The first four pairs belong to the head and are forwardly directed. They have large gnathobases that evidently crushed the food since no jaws are developed.

The **Crustacea** include lobsters, shrimps, crabs, barnacles, etc., and are predominantly marine animals; a smaller number inhabit fresh water, while a few kinds of crabs and the woodlice have invaded the land. They are characterized by the possession of two pairs of antennae followed by a pair of mandibles and at least five pairs of legs. In the higher forms the body segments are fixed in number and are grouped into two regions – the cephalothorax and abdomen. The appendages are for the most part specialized to perform a number of functions and are often of the biramous and gnathobasic type. The excretory organs are modified coelomoducts and are usually represented by green glands or shell glands. The genital apertures are located anteriorly, i.e., on the ninth post-oral segment in some cases, up to the fourteenth in others.

The **Chelicerata** include scorpions, king crabs, spiders, mites, ticks, sea-spiders and others and have the body divided into cephalothorax and abdomen. There are no antennae, these organs being replaced by prehensile chelicerae. True jaws are wanting and a varying number of the anterior limbs have developed gnathobases for breaking up the food. Four pairs of legs are present.

The **Onychophora** are in many ways intermediate between the Annelida or worms and the other Uniramia. They are represented by rather more than 70 species included in the genus *Peripatus* and its allies. Like many declining groups, they have a discontinuous geographical distribution and are found in warm countries in many parts of the world. Their relationship with arthropods is based on the presence of (1) paired limbs ending in claws; (2) respiration taking place by means of tracheae; (3) the haemocoelic body cavity; (4) a heart with paired ostia, and (5) the general character of the reproductive system. Annelidan characters include (1) segmentally repeated coelomoducts; (2) the structure of the eyes; (3) the rudimentary cephalization, only three head segments being present, and (4) the unstriated muscular body wall. Onychophora inhabit permanently damp localities and occur more especially beneath the bark of trees and underneath stones. The antennae, unlike those of other arthropods, are pre-antennae and arise from the first head segment while the second segment bears the jaws. The tracheal system has a non-segmental arrangement since the absence of hard sclerites has allowed an irregular distribution. The tracheae are very fine tubes, 2–3 μm in diameter, which arise in dense bundles from numerous flask-like pits in the integument.

The higher arthropods have a thicker and more rigid cuticle and consequently individual sclerites became developed so as to allow flexibility. The appendages for the same reason have acquired a jointed structure and are more complex than the lobe-like (lobopodial) appendages of the Onychophora.

The **Myriapoda** have a 5- or 6-segmented head bearing a single pair of antennae. The trunk is composed of numerous leg-bearing segments and is without differentiation into thorax and abdomen. The tracheal system is provided with segmentally repeated spiracles and the excretory organs are Malpighian tubules. These animals almost always hatch from the egg with a smaller number of trunk segments and limbs than are present when they are sexually mature. The addition of new segments takes place by subdivision of the penultimate segment. Myriapoda are divided into two chief groups – the Chilopoda, or centipedes, and the Diplopoda, or millipedes. The Symphyla and Pauropoda constitute two much smaller groups of some phylogenetic interest.

The Chilopoda have long, many-segmented antennae and the mouthparts comprise three pairs of appendages similar to those found in insects. The first pair of legs is modified into jaw-like poison claws and the gonopore is on the penultimate segment of the body.

The Diplopoda have short 7-segmented antennae and their mouthparts are more specialized than those of Chilopoda. The body segments are mostly grouped in pairs under each apparent tergum and the gonopore is on the third segment.

The Symphyla are small colourless Myriapoda with long, many-segmented antennae and mouthparts resembling those of Insecta. The trunk consists of 14 segments, each bearing a pair of append-ages and usually a pair of styli and protrusible vesicles. The gonopore is on the fourth trunk segment.

The Pauropoda are small, blind, soft-bodied forms, living in the soil or in other concealed habitats. They have large, branched antennae, mandibles, and one pair of maxillae. The last head segment bears no appendages and the eleven trunk segments are arranged like those of the Diplopods, with 9 or 10 pairs of legs and an anterior gonopore.

The **Insecta** (Hexapoda) show characters that ally them more closely with the Myriapoda than with any other arthropods. The division of the body into head, thorax and abdomen, the presence of only three pairs of legs and usually of two pairs of wings readily distinguish adult insects from other arthropods.

Reviewing the arthropods as a whole, one may first differentiate between the Uniramia and the others. The Uniramia, as their name implies, have appendages with a single branch or *ramus*, and their jaws bite with the tips. The Trilobita and Crustacea, on the other hand, have a characteristic biramous type of limb and in them and the Chelicerata the jaws are essentially gnathobasic structures derived from the proximal region of the limbs. The primitive Uniramia were probably soft-bodied forms with many similar lobopodial trunk appendages and little indication of a dif-ferentiated head. The Onychophora retained the thin cuticle and the many-legged lobopodial condition and did not evolve a head capsule, though they have pre-oral antennae and have developed sclerotized jaws on the second segment of the head. In the other Uniramia, definitive jaws (mandibles) evolved on the fourth cephalic segment, but their mode of action differed in the two main lines of descent. The Myriapodan groups evolved transversely biting mandibles, while in the more primitive insects these moved at first with a rotary, rolling action and only secondarily did a transverse type of movement arise (in the winged insects and their immediate ancestors). Increasing general sclerotization, with the development of a consolidated head capsule, seems to have taken place convergently in the Myriapod and the Insect groups, but whereas the Myriapods retained a large number of more or less undifferentiated trunk segments and many similar appendages, the insect trunk underwent division into thorax and abdomen, correlated with the emergence of a hexapod (six-legged) form of locomotion. Whether the hexapod condition arose separately in the four main groups of primitive wingless insects, or whether all are derived from a single primitive hexapod stock, remains uncertain. Manton, on whose detailed morphological and functional analysis the above account is largely based, considers that the hexapod condition arose separately in the different major insect groups. It follows from her conclusions that neither the phylum Arthropoda nor the comprehensive class Insecta can be regarded as natural monophyletic groups, though it should be mentioned that some

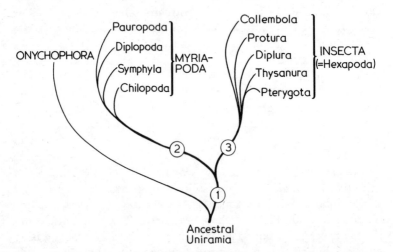

Fig. 93. Diagram to illustrate possible phylogeny of Uniramian arthropods

1, 2 and 3 denote evolution of functionally important characteristics: 1, jaws developing on fourth segment, soft, many-legged body, lobopodial limbs. 2, sclerotized head capsule, jointed, transversely-biting mandibles, one pair tentorial apodemes, many-legged trunk, lobopodial limbs. 3, sclerotized head capsule, unjointed rotary mandibles, two pairs tentorial apodemes, labium, many-legged trunk, lobopodial limbs (based on Manton)

morphologists are unwilling to admit the possibility of such large-scale evolutionary convergence as this interpretation requires.

THE ANCESTRY OF INSECTS

Attempts to trace the ancestry of insects have relied on evidence from embryology, comparative anatomy and palaeontology. Of these, the least informative so far has been palaeontology. The oldest known fossil insects are fragmentary remains of Collembola from the Middle Devonian rocks of Scotland. These belong to the species *Rhyniella praecursor* but unfortunately they tell us little of the origin of insects since the Collembola are generally agreed to be a somewhat specialized evolutionary side-branch. More generalized Apterygotes are not known until the Upper Carboniferous and Lower Permian (*Dasyleptus*, a representative of the purely fossil order Monura) or the Triassic (*Triassomachilis*). These must, however, have been preceded by much earlier Apterygotes, since fully evolved winged insects (Pterygota) are already known from the Upper Carboniferous. There are no fossils linking the early Apterygote insects with other arthropod groups.

In the absence of direct palaeontological evidence on the origin of insects, therefore, most theories have depended heavily on morphological data. Among the various hypotheses proposed, the most widely accepted until recently was the view that the insects shared a common ancestry with the Symphyla. Imms in particular has emphasized several structural features that are shared by the Diplura and the Symphyla and which include:

(a) many-segmented antennae, each segment except the last with intrinsic muscles – a character present in no other insects except the Collembola.

(b) absence of eyes and ocelli.

(c) mouthparts composed of mandibles, maxillae and labium.

(d) legs with five segments, the tarsi undivided and bearing paired claws.

(e) an abdomen with most of its segments bearing movable styles and eversible vesicles.

(f) a pair of short cerci receiving the opening of paired glands at their apices.

(g) the absence of an amnion and serosa in the developing embryo. The gonopore of the Symphyla is anterior, but this is a secondary condition and was, therefore, not thought to preclude a close relationship between the two groups.

More recently, the work of Manton on the detailed functional anatomy of the arthropod head, mouthparts and thorax, has cast doubt on the Symphylan theory of insect origins. As pointed out above (p. 214), Manton believes that jointed limbs and increasing sclerotization of the body and head capsule evolved independently in the myriapods and the hexapods and that even among the latter

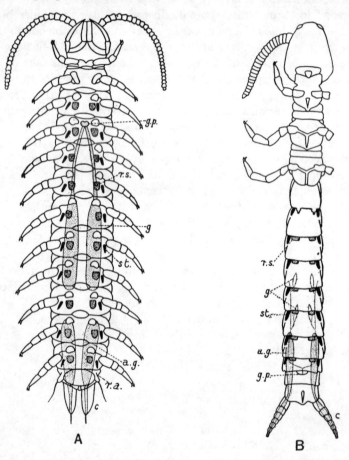

Fig. 94. Structural features of, A, Symphyla, and B, Diplura
a.g, posterior gland; *c*, cercus; *g*, gonad; *g.p*, gonopore; *r.a*, rudimentary appendage; *r.s*, protrusible vesicle or sac; *st*, stylus. (B *from* Silvestri)

the six-legged condition arose several times from a many-legged ancestral stock. A common progenitor for the Myriapods and the insects must therefore be sought at an earlier stage of Uniramian evolution which is not represented among living forms or in the known fossil record. Modern embryological evidence, especially that provided by the 'fate maps' of various arthropod groups (p. 114), has tended to support Manton's views without, however, establishing more precise indications of the origin of the insect groups.

MUTUAL RELATIONSHIPS

The four living Apterygote orders are not especially like one another structurally and if the hexapod condition really arose independently in each of them, they have no close phylogenetic links. The Diplura, Collembola and Protura, however, have mouthparts that are retracted into a cephalic pouch and for this reason are sometimes contrasted with the remaining Apterygote order, the Thysanura, which has normal external mouthparts, though it is not unlikely that the retracted (entognathous) condition also arose by convergence in the three orders that now show it. The Thysanura include *Lepisma* and its allies, which exhibit several features that suggest a close relationship with the more primitive Pterygote insects. This is shown by the structure of the tentorium, the presence of two main mandibular articulations, the appearance of a gonangulum in the ovipositor, and the presence of an amnion. Again, however, there is no palaeontological information to support this view of the origins of winged insects.

Among the most primitive winged insects are those contained in the fossil order Palaeodictyoptera (Fig. 95) whose remains are mostly found in rocks of Carboniferous and Permian age. The Palaeodictyoptera were characterized by:

(1) A rounded head bearing annulate antennae, specialized haustellate mouthparts, compound eyes and ocelli.

(2) The prothorax bearing lateral expansions resembling wing buds and the presence of similar outgrowths at the sides of the abdominal segments. It has been suggested that the ancestral insects had a series of such expansions all along the body that may have functioned as gliding planes. Of these, the meso- and meta-

thoracic pairs became greatly enlarged, acquired articulations with the thorax and ultimately developed into wings.

(3) The meso- and metathorax being subequal and bearing wings almost alike in form, size and venation and with broad basal attachments to the thorax; the venation more complete than in any other order and displaying an alternating sequence of convex and concave veins; the wing membrane strengthened by an irregular meshwork of veinlets or *archedictyon*; the wings held outstretched at rest and never folded back over the abdomen.

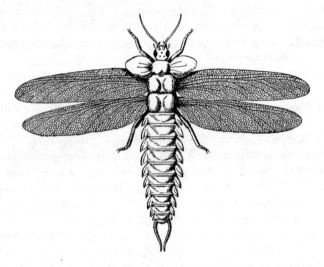

Fig. 95. *Stenodictya lobata* (Palaeodictyoptera)
(Reconstruction by Handlirsch)

(4) An elongate, 11-segmented abdomen bearing well-developed jointed cerci and a long ovipositor.

An inability to flex the wings over the abdomen while at rest seems to have been characteristic of several other early groups of Exopterygote insects and has led to them being classified together as the Palaeoptera, in contrast to the Neoptera, which includes all the remaining Pterygote orders. The only living Palaeoptera are the Ephemeroptera (mayflies), known first from the later Upper Carboniferous, and the Odonata (dragonflies), whose earliest representatives seem to be from the lower Permian, though they are closely related to an earlier fossil order, the Protodonata.

A major group of Neopteran insects is the complex of Orthopteroid orders. These include some purely fossil groups such as the Protorthoptera, together with nine living orders (Plecoptera, Grylloblattodea, Orthoptera, Phasmida, Dermaptera, Embioptera, Dictyoptera, Isoptera and Zoraptera). In general these have simple mandibulate mouthparts, a large anal lobe in the hind wing, cerci, many Malpighian tubules and several separate abdominal ganglia in the ventral nerve cord. Their mutual relationships are not easy to unravel though it seems likely that the Orthoptera and Phasmida are closely related, as are also the Dictyoptera and Isoptera. One order, the Grylloblattodea, is notable for its primitive features – it is virtually a 'living fossil' – while the small order Zoraptera forms a link between the Orthopteroid orders and the other main Exopterygote complex of Hemipteroid orders.

The most generalized Hemipteroid insects are the Psocoptera, known already from the Lower Permian and sharing several morphological peculiarities with the biting lice (Mallophaga) which, in turn, are closely related to the sucking lice or Siphunculata. The largest Hemipteroid order, the Hemiptera, is also thought to have been derived from early Psocopteran-like stock, a view based mainly on the venation of early Homoptera from the Permian. The affinities of the last Hemipteroid order, the Thysanoptera, are obscure.

A major step in insect evolution was the origin of the holometabolous life cycle, with its associated pupal instar. As a result of this, the larval and adult stages of Endopterygote insects have tended to diverge greatly from each other in structure and habits, reducing competition between them and increasing enormously the range of habitats colonized by the larva. Again, however, we have little precise information on the origin of the earliest Endopterygotes and the few theories on the subject are inconclusive. Viewed as a whole, though, the Endopterygota seem to show three lines of descent. The main evolutionary line, the Panorpoid complex, includes the orders Mecoptera, Lepidoptera, Trichoptera, Diptera and Siphonaptera. The common ancestor of these five groups may not have been very different from the most primitive living Mecoptera, such as *Merope*. Evolution of the wings has been mainly by reduction of the main veins and cross-veins. Primitive Lepidoptera and Trichoptera are very alike and almost certainly had a common ancestor, perhaps not very different from *Microp-*

terix. The Diptera are characterized by their narrow wing bases and the consequent reduction of Cu_2 and the anal veins. Their most primitive representatives are found among early types not unlike some of the living crane-flies and perhaps derived from 4-winged Mecopterous fossil forms, whose venation is very like that of existing primitive Nematocera, with characteristic narrow wing bases and correspondingly reduced anal areas. However, no annectant forms have been found linking the Mecoptera and Diptera, nor do the former show any tendency towards reduction of the hind wings.

The Neuroptera usually have a fuller venation than any of the Panorpoid groups and the larva of the Megaloptera is more primitive than any other, especially in its mouthparts. The Neuroptera must have diverged from the earliest Panorpoid stem and the Coleoptera which, though so very different as adults, have a fundamentally similar type of larva, may be thought to be derived from somewhere near the same point. This line lacks labial silk glands in the larva which therefore does not produce silk from its mouth though it sometimes does so from the anus. The Strepsiptera are traditionally placed near or in the Coleoptera, but their true relationships are not clear.

The Hymenoptera differ from the other Endopterygota in their peculiar wing venation and complete ovipositor, but their holo-metabolous metamorphosis suggests that they are derived from the same stock, probably also from ancestors of the Neuroptera.

Considered in broad, functional terms, the history of the insects thus shows a succession of major evolutionary steps – the nodal points of Hinton – each facilitating the successful radiation of the group into more diverse environments. The first of these was the development of mechanically effective forms of hexapod loco-motion, followed by the appearance of a tracheal system to allow efficient respiratory and humidity exchanges. This in turn was succeeded by the evolution of a relatively impermeable integument that permitted the exploitation of dry terrestrial environments. A fourth major step was the elaboration of thoracic paranotal lobes into wings adapted for flapping flight; while the last nodal point is represented by the origin of a pupal instar and the tendency to dissociate the nutritive and developmental functions of the larva from the reproductive and distributive capacities of the adult. Superimposed on these major changes have been other more restricted and specific innovations, often concerned with feeding, reproduction and communication, that have promoted the success of individual groups and so helped to make possible the immense variety of insect life known today.

7

Appendix on Literature

The following is a small selection from the very extensive literature on entomology.

General Works

BEIER, M. (Ed.) (1968→) Arthropoda. Insecta. In: *Handbuch der Zoologie.* (Eds. Helmcke, J.-G., Starck, D., & Wermuth, H.) Berlin: De Gruyter. 2. Auflage, Bd. 4., 2. Hälfte.

BORROR, D. J., DELONG, D. M. and TRIPLEHORN, C. A. (1976) *Introduction to the Study of Insects.* 4th Edn. New York: Holt, Rinehart and Winston. 864 pp.

BRUES, C. T., MELANDER, A. L. and CARPENTER, F. M., (1954) *Classification of Insects.* 2nd Edn. Cambridge, Mass.: Harvard University Press. 917 pp.

CANDY, D. J. and KILBY, B. A. (Eds.) (1975) *Insect Biochemistry and Function.* London: Chapman & Hall. 328 pp.

CHAPMAN, R. F. (1969) *The Insects: Structure and Function.* London: English Universities Press. 818 pp.

ESSIG, E. O. (1942) *College Entomology.* New York: Macmillan. 900 pp.

—— (1958) *Insects and Mites of Western North America.* 2nd Edn. New York: Macmillan. 1056 pp.

FROST, S. W. (1959) *Insect Life and Natural History.* New York: Dover. 526 pp.

GRANDI, G. (1951) *Introduzione alle Studio dell'Entomologia.* 2 vols. Bologna: Calderini.

GRASSÉ, P. P. (Ed.) (1949–75) *Traité de Zoologie. Insectes.* Vols. 8, 9, 10. Paris: Masson.

MACKERRAS, I. M. (Ed.) (1970, 1974) *The Insects of Australia.* Melbourne: C.S.I.R.O. 1029 pp., Suppl., 146 pp.

RICHARDS, O. W. and DAVIES, R. G. (1977) *Imms' General Testbook of Entomology.* 10th Edn., 2 vols. London: Chapman & Hall.

ROCKSTEIN, M. (Ed.) (1973–74) *The Physiology of Insecta.* 2nd Edn., 6 vols. New York: Academic Press.

SCHRÖDER, C. (Ed.) (1925–29) *Handbuch der Entomologie*. 3 vols. Jena: Fischer.

SÉGUY, S. (1967) Dictionnaire des termes techniques d'entomologie élémentaire. *Encycl. Ent.*, **41**: 465 pp.

SNODGRASS, R. E. (1935) *Principles of Insect Morphology*. New York: McGraw-Hill. 667 pp.

WEBER, H. (1933) *Lehrbuch der Entomologie*. Jena: Fischer. 726 pp. (Reprinted, Koenigstein, 1966).

—— (1974) *Grundriss der Insektenkunde*. 5th Edn. Stuttgart: Fischer. 640 pp. (revised by H. Weidner).

WIGGLESWORTH, V. B. (1972) *Principles of Insect Physiology*. 7th Edn. London: Chapman & Hall. 827 pp.

Special Aspects of Entomology

BARTON BROWNE, L. (Ed.) (1974) *Experimental Analysis of Insect Behaviour*. Berlin: Springer.

BIRCH, M. C. (Ed.) (1974) *Pheromones*. Amsterdam: North Holland 495 pp.

BRUES, C. T. (1946) *Insect Dietary*. Cambridge, Mass: Harvard University Press. 466 pp.

BUCHNER, P. (1965) *Endosymbiosis of Animals with Plant Micro-organisms*. New York: Interscience. 909 pp.

BUSNEL, R. G. (Ed.) (1963) *Acoustic Behaviour of Animals*. London & New York: Elsevier. 933 pp.

CARTER, W. (1973) *Insects in relation to Plant Disease*. New York: McGraw-Hill. 759 pp.

CLARK, L. R., GEIER, P. W., HUGHES, R. D. and MORRIS, R. F. (1967) *The Ecology of Insect Populations in Theory and Practice*. London: Chapman & Hall. 232 pp.

CLAUSEN, C. P. (1940) *Entomophagous Insects*. New York: McGraw-Hill. 688 pp.

COTT, H. B. (1957) *Adaptive Coloration in Animals*. London: Methuen. 508 pp.

COUNCE, S. J. and WADDINGTON, C. H. (Eds.) (1972) *Developmental Systems: Insects*. 2 vols. London & New York: Academic Press.

DANILEVSKII, A. S. (1965) *Photoperiodism and Seasonal Development of Insects*. Edinburgh & London: Oliver & Boyd. 282 pp.

DEBACH, P. (1964) *Biological Control of Insect Pests and Weeds*. London: Chapman & Hall. 844 pp.

DETHIER, V. G. (1963) *The Physiology of Insect Senses*. London: Methuen 266 pp.

EDNEY, E. B. (1977) *Water Balance in Land Arthropods*. Berlin: Springer. 282 pp.

EMDEN, H. F. van (Ed.) (1973) Insect/Plant Relationships. *Symp. R. ent. Soc. Lond.*, **6**: 215 pp.

ENGELMANN, F. (1970) *The Physiology of Insect Reproduction*. Oxford: Pergamon Press. 307 pp.

FELT, E. P. (1940) *Plant Galls and Gall Makers*. Ithaca, N.Y.: Cornstock. 364 pp.

FORD, E. B. (1964) *Ecological Genetics*. London: Methuen. 335 pp.

FRAENKEL, G. S. and GUNN, D. L. (1961) *The Orientation of Animals. Kineses, Taxes and Compass Reactions*. 2nd Edn. New York: Dover. 376 pp.

FRISCH, K. von (1967) *The Dance Language and Orientation of Bees*. Cambridge, Mass: Harvard University Press. 566 pp.

GILBERT, L. J. (Ed.) (1976) *The Juvenile Hormones*. New York: Plenum Press. 572 pp.

HAGAN, H. R. (1951) *Embryology of the Viviparous Insects*. New York: Ronald Press Co. 472 pp.

HEPBURN, H. R. (Ed.) (1976) *The Insect Integument*. Amsterdam: Elsevier. 571 pp.

HERING, E. M. (1951) *Biology of the Leaf Miners*. The Hague: Junk. 420 pp.

HORRIDGE, G. A. (Ed.) (1974) *The Compound Eye and Vision of Insects*. Oxford: Clarendon Press. 595 pp.

JACOBSON, M. (1972) *Insect Sex Pheromones*. New York: Academic Press. 382 pp.

JOHANNSEN, O. A. and BUTT, F. H. (1941) *Embryology of Insects and Myriapods*. New York: McGraw-Hill. 462 pp.

JOHNSON, C. G. (1969) *Migration and Dispersal of Insects by Flight*. London: Chapman & Hall. 763 pp.

LAWRENCE, P. A. (Ed.) (1976) Insect Development. *Symp. R. ent. Soc. Lond.*, **8**: 240 pp.

LEES, A. D. (1955) *The Physiology of Diapause in Arthropods*. Cambridge University Press. 150 pp.

MAYR, E. (1969) *Principles of Systematic Zoology*. New York: McGraw-Hill. 428 pp.

NEVILLE, A. C. (1975) *Biology of the Arthropod Cuticle*. Berlin: Springer. 448 pp.

NOVÁK, J. A. (1975) *Insect Hormones*. 2nd Edn. London: Chapman & Hall. 624 pp.

PETERSON, A. (1960) *Larvae of Insects. An Introduction to Nearctic Species*, 4th Edn. Columbus, Ohio: The Author. 416 pp.

PRINGLE, J. W. S. (1957) *Insect Flight*. Cambridge University Press. 133 pp.

RAINEY, R. C. (Ed.) (1975) Insect Flight. *Symp. R. ent. Soc. Lond.*, 7: 287 pp.

RODRIGUEZ, J. G. (Ed.) (1973) *Insect and Mite Nutrition.* Amsterdam: North Holland. 717 pp.

SALT, G. (1970) *The Cellular Defence Reactions of Insects.* Cambridge University Press. 117 pp.

SAUNDERS, D. S. (1975) *Insect Clocks.* Oxford: Pergamon. 312 pp.

SCHMIDT, G. H. (Ed.) (1974) *Sozialpolymorphismus bei Insekten.* Stuttgart: Wissenschaftliche Verlagsgesellschaft. 974 pp.

SMITH, K. G. V. (Ed.) (1973) *Insects of Medical Importance.* London: British Museum (Nat. Hist.).

SNODGRASS, R. E. (1954) Insect metamorphosis. *Smithson. misc. Collns,* **122** (9): 124 pp.

STEINHAUS, E. A. (1963) *Insect Pathology: an advanced treatise.* New York: McGraw-Hill. 2 vols.

STRAUSFELD, N. J. (1976) *Atlas of an Insect Brain.* Berlin: Springer. 230 pp.

TINBERGEN, N. (1951) *The Study of Instinct.* Oxford: Clarendon Press. 228 pp.

URSPRUNG, H. and NÖTHIGER, R. (Ed.) (1972) *The Biology of Imaginal Disks.* Berlin: Springer. 172 pp.

USHERWOOD, P. N. R. (Ed.) (1975) *Insect Muscle.* London & New York: Academic Press. 621 pp.

VARLEY, G. C., GRADWELL, G. R. and HASSELL, M. P. (1973) *Insect Population Ecology. An Analytical Approach.* Oxford: Blackwell. 212 pp.

WESENBERG-LUND, C. (1943) *Biologie der Süsswasserinsekten.* Copenhagen: Nordisk Forlag. 682 pp.

WHITE, M. J. D. (1973) *Animal Cytology and Evolution.* 3rd Edn. Cambridge University Press. 961 pp.

WIGGLESWORTH, V. B. (1970) *Insect Hormones.* Edinburgh & London: Oliver & Boyd. 159 pp.

WILSON, E. O. (1971) *The Insect Societies.* Cambridge, Mass: Harvard University Press. 548 pp.

For reviews of recent work in various aspects of entomology see the annual volumes of *Advances in Insect Physiology* and *Annual Review of Entomology.*

Works on the Orders of Insects

1. *Thysanura*

DELANY, M. J. (1954) Thysanura and Diplura. *R. ent. Soc. Lond. Handb. Ident. Brit. Ins.*, **1** (2): 7 pp.

JANETSCHEK, H. (1954) Ueber Felsenspringer der Mittelmeerländer. *Eos*, 30: 163–314.

LINDSAY, E. (1940) The biology of the silverfish *Ctenolepisma longicaudata* Esch., with particular reference to its feeding habits. *Proc. R. Soc. Vict.* (*N.S.*), 52: 35–83.

PALISSA, A. (1964) Apterygota – Urinsekten. *Tierwelt Mitteleuropas*, 4 (I): 407 pp.

REMINGTON, C. L. (1954) The suprageneric classification of the order Thysanura (Insecta). *Ann. ent. Soc. Am.*, 47: 277–86.

SAHRHAGE, D. (1953) Oekologische Untersuchungen an *Thermobia domestica* (Packard) und *Lepisma saccharina* L., *Z. wiss. Zool.*, 157: 77–168.

WYGODZINSKY, P. (1972) A review of the silverfish (Lepismatidae, Thysanura) of the United States and Caribbean area. *Am. Mus. Novit.*, 2481: 1–26.

2. *Diplura*

CONDÉ, B. (1956) Matériaux pour une monographie des Diploures Campodéidés. *Mem. Mus. nat. Hist., Paris*, 12 (1955): 1–201.

GYGER, H. (1960) Untersuchungen zur postembryonalen Entwicklung von *Dipljapyx humberti* (Grassi). *Verh. naturf. Ges. Basel*, 71: 29–95.

MARTEN, W. (1939) Zur Kenntnis von *Campodea*. *Z. Morph. Ökol. Tiere*, 36: 40–88.

PACLT, J. (1957) Diplura. *Genera Insectorum*, Fasc. 212: 123 pp.

SMITH, L. M. (1961) Japygidae of North America. 8. *Ann. ent. Soc. Am.*, 54, 437–41.

See also Delany, Paclt and Palissa under Thysanura and Collembola.

3. *Protura*

JANETSCHEK, H. (1970) Ordnung Protura (Beintastler). In: *Handbuch der Zoologie*, Bd. 4, 2. Hälfte, Lfg. 14, Beitr. 3: 1–58.

RAW, F. (1956) The abundance and distribution of Protura in grassland. *J. Anim. Ecol.*, 25: 15–21.

TUXEN, S. L. (1949) Uber den Lebenszyklus und die postembryonale Entwicklung zweier dänischer Proturengattungen. *K. danske Vidensk. Selsk. Skr.*, 6: 49 pp.

—— (1964) *The Protura*. Paris: Hermann. 360 pp.

YIN, W. (1968) Studies on Chinese Protura, 2. *Acta ent. sin.*, 2: 26–34.

See also Palissa under Thysanura.

4. *Collembola*

AGRELL, I. (1949) Studies on the postembryonic development of Collemboles. *Ark. Zool.*, 41A (12): 1–35.

BUTCHER, J. W., SNIDER, R. and SNIDER, R. J. (1971) Bioecology of edaphic Collembola and Arachnida. *A. Rev. Ent.*, **16**: 240–88.

CHRISTIANSEN, K. A. (1964) Bionomics of Collembola. *A.Rev. Ent.*, **9**: 147–78.

GISIN, H. (1960) *Collembolenfauna Europas*. Geneva: Natural History Museum. 312 pp.

PACLT, J. (1956) *Biologie der primär flügellosen Insekten*. Jena: Fischer. 258 pp.

SALMON, J. T. (1964) An index to the Collembola. *Bull. R. Soc. New Zealand*, **7**: 644 pp.

SCOTT, H. G. (1961) Collembola. Pictorial keys to the Nearctic genera. *Ann. ent. Soc. Am.*, **54**: 104–13.

STACH, J. (1947–63) The Apterygotan fauna of Poland in relation to the world-fauna of this group of insects. *Acta monograph. Mus. hist. nat. Polska Akad. Umiej.*, **1947**: 488 pp.; **1949**: 341 pp.; **1949**: 122 pp.; **1951**: 100 pp.; *Polsk. Akad. Nauk. Cracow*, **1954**: 219 pp.; **1956**: 287 pp.; **1957**: 113 pp.; **1960**: 151 pp.; **1963**: 126 pp.

5. *Ephemeroptera*

BERNER, L. (1959) Tabular summary of the biology of North American mayfly nymphs. *Bull. Fla. St. Mus.*, **4**: 1–58.

DEMOULIN, G. (1958) Nouveau schéma de classification des Archodonates et des Ephéméroptères. *Bull. Inst. r. Sci. Belg.*, **34** (27): 1–19.

EDMUNDS, G. F. (1972) Biogeography and evolution of Ephemeroptera. *A. Rev. Ent.*, **17**: 21–42.

IDE, F. P. (1935) Post-embryological development of Ephemeroptera (mayflies), external characters only. *Can. J. Res.*, **12**: 433–78.

ILLIES, J. (1968) Ephemeroptera (Eintagsfliegen). In: Helmcke, J.-G., Starck, D., & Wermuth, H. (Eds.) *Handbuch der Zoologie*, **4** (2) 63 pp.

KIMMINS, D. E. (1972) A revised key to the adults of the British species of Ephemeroptera with notes on their ecology. *Scient. Publs Freshwater biol. Ass.*, (2nd revised edn.) **15**: 75 pp.

MACAN, T. T. (1970) A key to the nymphs of British species of Ephemeroptera with notes on their ecology. *Scient. Publs Freshwater biol. Ass.* (2nd revised edn.). **20**: 68 pp.

NEEDHAM, J. G., TRAVER, J. R. and HSU, Y (1935) *The Biology of Mayflies*. Ithaca, N.Y.: Comstock. 759 pp.

6. *Odonata*

CORBET, P. S. (1962) *A Biology of Dragonflies*. London: Witherby. 247 pp.

CORBET, P. S., LONGFIELD, C. and MOORE, N. W. (1960) *Dragonflies*. London: Collins. 260 pp.

FRASER, F. C. (1956) Odonata. *R. ent. Soc. Handb. Ident. Brit. Ins.*, **1** (10): 49 pp.

—— (1957) *A Reclassification of the Order Odonata.* Sydney: R. Zool. Soc., N.S.W. 133 pp.

GARDNER, A. E. (1954) A key to the larvae of the British Odonata. *Ent. Gaz.*, **5**: 157–71; 193–213.

NEEDHAM, J. G. and WESTFALL, M. J. (1954) *A Manual of the Dragonflies of North America (Anisoptera).* Berkeley University of California Press.

ST. QUENTIN, D. and BEIER, M. (1968) Odonata (Libellen). In: *Handbuch der Zoologie* (Eds. Helmcke, J.-G., Starck, D., & Wermuth, H.) **4** (2): Lfg. 3., 39 pp.

SNODGRASS, R. E. (1954) The dragonfly larva. *Smithson. misc. Collns*, **123** (2): 38 pp.

7. *Plecoptera*

BRINCK, P. (1949) Studies on Swedish stoneflies. *Opusc. ent., Suppl.*, **11**: 250 pp.

CLAASSEN, P. W. (1931) *Plecoptera Nymphs of North America.* Springfield: Thomas Say Foundation. 199 pp.

HYNES, H. B. N. (1967) A key to the adults and nymphs of British stoneflies (Plecoptera). *Scient. Publs Freshwater biol. Ass.*, **17**: 1–90.

—— (1976) Biology of the Plecoptera. *A. Rev. Ent.*, **21**: 135–53.

ILLIES, J. (1965) Phylogeny and zoogeography of the Plecoptera. *A. Rev. Ent.*, **10**: 117–40.

—— (1966) Katalog der rezenten Plecoptera. *Tierreich*, **82**: 632 pp.

NEEDHAM, J. G. and CLAASSEN, P. W. (1925) *A Monograph of the Plecoptera or Stoneflies of America north of Mexico.* Lafayette: Thomas Say Foundation. 197 pp.

ZWICK, P. (1973) Insecta: Plecoptera. Phylogenetisches System und Katalog. *Tierreich*, **94**: 465 pp.

8. *Grylloblattodea*

GURNEY, A. B. (1961) Further advances in the taxonomy and distribution of the Grylloblattidae. *Proc. biol. Soc. Washington*, **74**: 67–76.

KAMP, J. W. (1970) The cavernicolous Grylloblattodea of the western United States (i). *Ann. spéléol.*, **25**: 223–30.

WALKER, E. M. (1937) *Grylloblatta*, a living fossil. *Trans. R. Soc. Can.*, **31**: 1–10.

9. *Orthoptera*

BEIER, M. (1972) Saltatoria (Grillen und Heuschrecken). In: *Handbuch der Zoologie*, (Eds. Helmcke, J.-G., Starck, D., & Wermuth, H.) **4** (2), Lfg. **17**: 1–217.

BLATCHLEY, W. S. (1920) *The Orthoptera of North Eastern America.* Indianapolis: Nature Publishing Co., 784 pp.

CHOPARD, L. (1938) La biologie des Orthoptères. *Encycl. ent.*, **20**: 541 pp.

—— (1951) Orthopteroïdes. *Faune de France*, **56**: 359 pp.

DIRSH, V. M. (1975) *Classification of the Acridomorphoid Insects.* Faringdon: Classey, 184 pp.

HARZ, K. (1969, 1975) *Die Orthopteren Europas.* 2 vols. The Hague: Junk.

PIERCE, G. W. (1948) *The Songs of Insects.* Cambridge, Mass.: Harvard University Press, 329 pp.

RAGGE, D. R. (1965) *Grasshoppers, Crickets and Cockroaches of the British Isles.* London: Warne, 299 pp. (Supplement: *Ent. Gaz.*, **24**: 227–45, 1973).

RICHARDS, O. W. and WALOFF, N. (1954) Studies on the biology and population dynamics of British grasshoppers. *Anti-Locust Bull.*, **17**: 182 pp.

UVAROV, B. P. (1966, 1977) *Grasshoppers and Locusts. A handbook of general acridology.* Cambridge University Press, 2 vols: 481 pp., 597 pp.

10. *Phasmida*

BEIER, M. (1957) Arthropoda. Insecta. Orthopteroidea: Ordnung Cheleutoptera Crampton 1915 (Phasmida Leach 1815) In: Bronn's *Klassen u. Ordnungen des Tiere*, **5**, Abt. iii (6, 9): 305–454.

—— (1968) Phasmida (Stab- oder Gespenstheuschrecken). In: *Handbuch der Zoologie* (Eds. Helmcke, J.-G., Starck, D., & Wermuth, H. **4** (2) Lfg. 6: 56 pp.

KEY, K. H. L. (1957) Kentromorphic phases in three species of Phasmatodea. *Aust. J. Zool.*, **5**: 247–84.

11. *Dermaptera*

BEIER, M. (1958) Dermaptera. In: Bronn's *Klassen u. Ordnungen des Tierreichs*, **5** (Buch 6, Lfg. 3): 455–585.

GILES, E. T. (1963) The comparative external morphology and affinities of the Dermaptera. *Trans. R. ent. Soc. Lond.*, **115**: 95–164.

GÜNTHER, K. and HERTER, K. (1974) 11. Ordnung Dermaptera (Ohrwürmer). In: *Handbuch der Zoologie* (Eds. Helmcke, J.-G., Starck, D., & Wermuth, H.) **4** (2), Lfg. 23: 1–158.

NAKATA, S. and MAA, T. C. (1974) A review of the parasitic earwigs (Dermaptera: Arixenina, Hemimerina). *Pacific Insects*, **16**: 307–74.

POPHAM, E. J. (1965) The functional morphology of the reproductive organs of the common earwig (Forficula auricularia) and other Dermaptera with reference to the natural classification of the order. *J. Zool.*, **146**: 1–43.

STEINMANN, H. (1975) Suprageneric classification of Dermaptera. *Acta zool. Acad. Sci. hung.*, **21**: 195–220.

12. *Embioptera*

DAVIS, C. (1940) Family classification of the order Embioptera. *Ann. ent. Soc. Am.*, **33**: 677–82.

KALTENBACH, A. (1968) Embiodea (Spinnfüsser). In: *Handbuch der Zoologie* (Eds. Helmcke, J.-G., Starck, D., & Wermuth, H.), **4** (2), 2/8: 29 pp.

ROSS, E. S. (1944) A revision of the Embioptera or web-spinners of the New World. *Proc. U.S. nat. Mus.*, **94**: 401–504.

—— (1966) The Embioptera of Europe and the Mediterranean region. *Bull. Brit. Mus. (Nat. Hist), Entomol.*, **17** (7): 273–326.

—— (1970) Biosystematics of the Embioptera. *A. Rev. Ent.*, **15**: 157–72.

13. *Dictyoptera*

BEIER, M. (1968) Mantodea (Fangheuschrecken). In: *Handbuch der Zoologie* (Eds. Helmcke, J.-G., Starck, D., & Wermuth, H.) **4** (2), Lfg. 4: 47 pp.

—— (1974) 13. Ordnung Blattariae (Schaben). In: *Handbuch der Zoologie*, Helmcke, J.-G., Starck, D., & Wermuth, H. **4** (2), Lfg. 23: 127 pp.

GUTHRIE, D. M. and TINDALL, A. R. (1968) *The Biology of the Cockroach.* London: Arnold. 416 pp.

MCKITTRICK, F. A. (1964) Evolutionary studies of cockroaches. *Mem. Cornell Univ. agr. Exp. Sta.*, **389**, 197 pp.

PRINCIS, K. (1966) Zur Systematik der Blattaria. *Eos*, **36**: 427–49.

ROTH, L. M. (1970) Evolution and taxonomic significance of reproduction in Blattaria. *A. Rev. Ent.*, **15**: 75–96.

ROTH, L. M. and WILLIS, E. R. (1960) The biotic association of cockroaches. *Smithson. misc. Collns*, **141**: 470 pp.

See also works by Chopard (1938, 1951) and Ragge (1965) under Orthoptera.

14. *Isoptera*

HARRIS, W. V. (1971) *Termites, their recognition and control.* London: Longman. 2nd Edn., 187 pp.

HOWSE, P. E. (1970) *Termites: a study in social behaviour.* London: Hutchinson. 150 pp.

KRISHNA, K. and WEESNER, F. M. (1969–70) *Biology of Termites.* New York: Academic Press. 2 vols.

LEE, K. E. and WOOD, T. G. (1971) *Termites and Soils.* London & New York Academic Press. 252 pp.

SANDS, W. A. (1972) The soldierless termites of Africa (Isoptera: Termitidae). *Bull. Br. Mus. Nat. Hist. Suppl. (Ent.)*, **18**: 1–244.

SNYDER, T. E. (1949) Catalog of the termites (Isoptera) of the world. *Smithson. misc. Collns*, **112**: 1–490.

WEIDNER, H. (1970) Isoptera (Termiten). In: *Handbuch der Zoologie,* (Eds. Helmcke, J.-G., Starck, D., & Wermuth, H.) **4** (2), Lfg. 13: 147 pp.

15. *Zoraptera*

WEIDNER, H. (1969) Die Ordnung Zoraptera oder Bodenläuse. *Ent. Z.,* **79**: 29–51.

16. *Psocoptera*

BADONNEL, A. (1934) Recherches sur l'anatomie des Psoques. *Bull. biol. Fr. Belg., Suppl.,* **18**: 1–241.

NEW, T. R. (1974) Psocoptera. *R. ent. Soc. Handb. Ident. Brit. Ins.,* **1** (7): 102 pp.

SMITHERS, C. N. (1972) The classification and phylogeny of the Psocoptera. *Mem. Aust. Mus. Sydney,* **14**: 351 pp.

WEIDNER, H. (1972) 16. Ordnung Copeognatha (Staubläuse). In: *Handbuch der Zoologie* (Eds. Helmcke, J.-G., Starck, D., & Wermuth, H.) **4** (2), Lfg. 18: 1–94.

17. *Mallophaga*

CLAY, T. (1951) An introduction to the classification of the avian Ischnocera (Mallophaga), Part 1. *Trans. R. ent. Soc. Lond.,* **102**: 171–94.

—— (1970) The Amblycera (Phthiraptera: Insecta). *Bull. Brit. Mus. nat. Hist. (Ent.),* **25**: 73–98.

EICHLER, W. (1963) Mallophaga. In: Bronn's *Klassen u. Ordnungen des Tierreichs,* **5** (3), 7(b): 1–290.

HOPKINS, G. H. E. and CLAY, T. (1952) *A Check List of the Genera and Species of Mallophaga.* London: British Museum (Nat. Hist.) 362 pp.

KÉLER, S. von (1969) 17. Ordnung Mallophaga (Federlinge und Haarlinge). In: *Handbuch der Zoologie* (Eds. Helmcke, J.-G., Starck, D., & Wermuth, H.) **4** (2), Lfg. 10: 1–72.

See also Hopkins (1949) under Siphunculata.

18. *Siphunculata*

BUXTON, P. A. (1947) *The Louse. An account of the lice which infest man, their medical importance and control.* London: Arnold. 2nd Edn., 164 pp.

FERRIS, G. F. (1951) The sucking lice. *Mem. Pacific Coast ent. Soc.,* **1**: 1–320.

HOPKINS, G. H. E. (1949) The host-associations of the lice of mammals. *Proc. zool. Soc. Lond.,* **119**: 387–604.

19. *Hemiptera*

BALACHOWSKY, A. (1937–50) Les cochenilles de France, d'Europe, du nord de l'Afrique et du bassin méditerranéan, I–V. *Actualités Sci. et Industr.*, **526**: 68 pp.; **564**: 59 pp.; **784**: 111 pp.; **1054**: 154 pp.; **1087**: 163 pp.

BEIRNE, B. P. (1956) Leafhoppers (Homoptera Cicadellidae) of Canada and Alaska. *Can. Ent.* **88**, *Suppl.* 2: 180 pp.

BLATCHLEY, W. S. (1926) *Heteroptera of Eastern North America*. Indianapolis: Nature Publishing Co. 1116 pp.

BÖRNER, C. (1952) Europae centralis aphides. Die Blattlause Mittel-europas. Namen, Synonyme, Wirtspflanzen, Generationszyklen. *Schrift. Thüring. Land. Heilpflanz. Weimar*, **4** (3): 484 pp.

CHINA, W. E. and MILLER, N. C. E. (1959) Check-list and keys to the families and subfamilies of the Hemiptera-Heteroptera. *Bull. Br. Mus. nat. Hist. (Ent.)*, **8** (1): 45 pp.

COBBEN, R. H. (1968) Evolutionary trends in Heteroptera. Part 1. Eggs, architecture of the shell, gross embryology and eclosion. *Meded. Lab. Ent., Wageningen*, **151**: 1–475.

DELONG, D. M. (1971) The bionomics of leafhoppers. *A. Rev. Ent.*, **16**: 179–210.

EVANS, J. W. (1963) The phylogeny of the Homoptera. *A. Rev. Ent.*, **8**: 77–94.

FERRIS, G. F. (1937–55) *An Atlas of the Scale Insects of North America*. Stanford, California: Stanford University Press. 6 vols.

GOODCHILD, A. J. P. (1966) Evolution of the alimentary canal in the Hemiptera. *Biol. Rev.*, **41**: 97–140.

HODKINSON, I. D. (1974) The biology of the Psylloidea (Homoptera): a review. *Bull. ent. Res.*, **64**: 325–39.

JORDAN, K. H. C. (1972) 20. Heteroptera (Wanzen). In: *Handbuch der Zoologie* (Eds. Helmcke, J.-G., Starck, D., & Wermuth, H.) **4** (2), Lfg. 16: 113 pp.

KENNEDY, J. S. and STROYAN, H. L. G. (1959) Biology of Aphids. *A. Rev. Ent.*, **4**: 139–60.

KRAMER, S. (1950) The morphology and phylogeny of Auchenorrhyn-chous Homoptera (Insecta). *Illinois biol. Monogr.*, **20** (4): 111 pp.

LEES, A. D. (1966) The control of polymorphism in aphids. *Adv. Insect Physiol.*, **3**: 207–77.

LEQUESNE, W. J. (1960–69) Hemiptera Cicadomorpha, Hemiptera Ful-goromorpha. *R. ent. Soc. Handb. Ident. Brit. Insects*, **II** (2a): 1–64; **II** (2b): 65–148; **II** (3): 1–68.

METCALF, Z. P. (1927–66) *General Catalogue of the Hemiptera* (in many parts, with contributions by W. D. Funkhouser, V. Wade, *et al.*).

NAST, J. (1972) *Palaearctic Auchenorrhyncha (Homoptera). An annotated check list*. Warsaw: Polish Science Publications. 350 pp.

NIELSON, M. W. (1968) The leafhopper vectors of phytopathogenic viruses (Homoptera, Cicadellidae): taxonomy, biology and virus transmission. *Tech. Bull. U.S. Dept. Agric.*, **1382**: 386 pp.

OMAN, P. W. (1949) The Nearctic leafhoppers (Homoptera: Cicadellidae). A generic classification and check list. *Mem. ent. Soc. Washington*, **3**: 253 pp.

PARSONS, M. C. (1964) The origin and development of the Hemipteran cranium. *Can. J. Zool.*, **42**: 409–32.

PESSON, P. (1951) Ordre des Homoptères. In: *Traité de Zoologie* Ed. Grassé, P. P. **10** (2): 1390–656.

POISSON, R. (1951) Ordre des Hétéroptères. In: *Traité de Zoologie* Ed. Grassé, P. P. **10** (2): 1657–803.

RIBAUT, H. (1936, 1952). Homoptères Auchénorrhynques, I, II. *Faune de France*, **31**: 228 pp., **57**: 474 pp.

RIS LAMBERS, D. H. (1938–53) Contributions to a monograph of the Aphididae of Europe, I–V. *Temminckia*, **3**: 1–43; **4**: 1–134; **7**: 173–319; **8**: 182–323; **9**: 1–176.

SAMPSON, W. W. (1943) A generic synopsis of the Hemipterous super-family Aleyrodoidea. *Entomologica am.*, **23**: 173–223.

SOUTHWOOD, T. R. E. and LESTON, D. (1959) *Land and Water Bugs of the British Isles.* London: Warne. 436 pp.

STICHEL, W. (1955–62) *Illustrierte Bestimmungstabellen der Wanzen. II. Europa.* Berlin: Buchdruckerei Erich Pröh. 4 vols.

THERON, J. G. (1958) Comparative studies on the morphology of male scale insects. (Hemiptera: Coccoidea). *Ann. Univ. Stellenbosch*, **34** (A) (1): 1–71.

TUTHILL, L. D. (1943) The Psyllids of America north of Mexico (Psyllidae: Homoptera) (Subfamilies Psyllinae & Triozinae). *Iowa St. Coll. J. Sci.*, **17**: 443–667.

USINGER, R. L., WYGODZINSKY, P., and RYCKMAN, R. E. (1966) The biosystematics of Triatominae. *A. Rev. Ent.*, **11**: 309–30.

WEBER, H. (1930) *Biologie der Hemipteren.* Berlin: Springer. 543 pp.

20. Thysanoptera

JACOT-GUILLARMOD, C. F. (1970–74) Catalogue of the Thysanoptera of the World. I–III. *Ann. Cape Prov. Mus. nat. Hist.*, **7**: 1–515.

LEWIS, T. (1973) *Thrips, their Biology, Ecology and Economic Importance.* London: Academic Press. 350 pp.

MOUND, L. A., MORISON, G. D., PITKIN, B. R. and PALMER, J. M. (1977) Thysanoptera. *R. ent. Soc. Handb. Ident. Brit. Ins.*, **1** (1): 79 pp.

PRIESNER, H. (1964) Ordnung Thysanoptera (Fransenflügler, Thripse). *Bestimmungsbucher zur Bodenfauna Europas.* Lfg. **2**: 242 pp.

—— (1968) Thysanoptera (Physapoda, Blasenfusser). In: *Handbuch der*

Zoologie (Eds. Helmcke, J.-G., Starck, D., & Wermuth, H.) **4** (2) Lfg. 5: 32 pp.

STANNARD, L. J. (1968) The Thrips, or Thysanoptera, of Illinois. *Bull. Ill. nat. Hist. Surv.*, **29** (4): 552 pp.

21. *Neuroptera*

CARPENTER, F. M. (1936) Revision of the Nearctic Raphidioidea (recent and fossil). *Proc. Am. Acad. Arts and Sciences*, **71**: 89–157.

—— (1940). A revision of the Nearctic Hemerobiidae, etc. *Ibid.*, **74**: 193–280.

DAVIS, K. C. (1903) Sialididae of North and South America. *Bull. N.Y. Mus.*, **58**: 442–86.

KILLINGTON, F. J. (1930) *A monograph of the British Neuroptera*. London, Ray Soc. **1** and **2**.

RIEK, E. F. (1970) Megaloptera. Neuroptera, In: *The insects of Australia*, chap. 28 and 29, 405–71; 472–94, Melbourne University Press.

WHEELER, W. M. (1930). *Demons of the dust*. London: Kegan Paul.

22. *Coleoptera*

BERTRAND, H. (1972) *Larves et nymphes des Coléoptères du globe.* Abbeville: Imprimerie Paillart.

BÖVING, A. G. and CRAIGHEAD, F. C. (1931) Larvae of Coleoptera. *Entomologia Am.*, **11**: 1–351.

CROWSON, R. A. (1955) *The Classification of the Families of British Coleoptera*. London: Nathaniel Lloyd.

FREUDE, H., HARDE, K. W. and LOHSE, G. A. (1964–74) *Die Käfer Mitteleuropas*, vols. **1–5, 7–9**. Krefeld: Goecke & Evers.

JEANNEL, R. and PAULIAN, R. (1949). Coléoptères. In: *Traité de Zoologie*, (Ed. Grassé, P. P.) **9**: 771–1077. Paris.

LENGERKEN, H. von (1954). *Die Brutfürsorge- und Brutpflegeinstinkte der Käfer*. Leipzig: Akademische Verlagsgesellschaft. 2nd Edn.

MEIXNER, J. (1935) Coleopteroidea. In: *Handbuch der Zoologie* (Ed. Kükenthal, W.) Bd. **4**: Insecta 2, Lief. 3–5. Berlin.

23. *Strepsiptera*

BOHART, R. M. (1941). A revision of the Strepsiptera with special reference to the species of North America. *Univ. Calif. Publ. Ent.*, **7**: 91–160.

KINZELBACH, R.K. (1971) Morphologische Befunde an den Fächerflüglern und ihre phylogenetische Bedeutung. (Insecta: Stepsiptera), *Zoologia*, **119**: 1–256.

PIERCE, W. D. (1909). A monographic revision of the twisted winged insects comprising the order Strepsiptera Kirby. *Bull. U.S. nat. Mus.*, **66**: xii + 232 pp.

ULRICH, W. (1943). Die Mengeiden (Mengenillini) und die Phylogenie der Strepsipteren. *Z. Parasitenk.*, **13**: 62–101.

24. Mecoptera

CARPENTER, F. M. (1931) Revision of the Nearctic Mecoptera. *Bull. Mus. comp. Zool.*, **72**: 205–77.

HOBBY, B. M. and KILLINGTON, F. J. (1934). The feeding habits of British Mecoptera, with a synopsis of the British species. *Trans. Soc. Brit. Ent.*, **1**: 39–49.

POTTER, E. M. (1938) The internal anatomy of the order Mecoptera. *Trans. R. ent. Soc. Lond.*, **87**: 467–501.

25. Siphonaptera

HOLLAND, G. P. (1949) The Siphonaptera of Canada. *Tech. Bull. Dep. Agric. Canada*, **70**: 306 pp.

ROTHSCHILD, M. and HOPKINS, G. H. E. (1953–71). *An Illustrated Catalogue of the Rothschild Collection of Fleas, etc.* **1–5**. London, British Museum.

SMIT, F. G. A. M. (1957) Siphonaptera. *R. ent. Soc. Hndb. Ident. Brit. Ins.*, **1**, Pt. 16.

26. Diptera

BUXTON, P. A. (1955) *The Natural History of the Tsetse flies*, London: School of Hygiene and Tropical Medicine. Mem. 10, 816 pp.

CRAMPTON, G. C. *et al.* (1942) Guide to the insects of Connecticut. Part VI. The Diptera or true flies of Connecticut. First fascicle. *Bull. Conn. geol. nat. Hist. Surv.*, **64**: pp. x + 509.

CURRAN, C. H. (1934) *The Families and Genera of North American Diptera.* New York: The Author.

HENDEL, F. (1936–7). Diptera. In: *Handbuch der Zoologie*, (Ed. Kükenthal, W.) Bd. **4**: Insecta 2, Lf. 8–11. Berlin.

HENNIG, W. (1948–52) *Die Larvenformen der Diptera*, **1–3**. Berlin: Akademie Verlag.

LINDNER, W. (Ed.) (1924–77) *Die Fliegen der paläarktischen Region.* (incomplete) Stuttgart: Schweitzerbartsch.

LUNDBECK, W. (1907–27). *Diptera Danica*, **1–7**. Copenhagen (unfinished): Verlagsbuchhandlung.

PANTEL, J. (1910). Recherches sur les Diptères à larves entomobies. *La Cellule*, **26**: 25–216.

VERRALL, G. H. and COLLIN, J. E. (1901–61). *British Flies*, **8, 5, 6**. London: Gurney & Jackson; Cambridge: University Press.

27. *Lepidoptera*

BOURGOGNE, J. (1951) Lépidoptères. In: *Traité de Zoologie*, (Ed. Grassé, P. P.) **10**, fasc. 1. Paris.

CLARK, A. H. (1932) The butterflies of the district of Columbia and vicinity. *Smithson. Inst. U.S. Nat. Mus. Bull.*, **137**: ix + 337 pp.

FORBES, W. T. M. (1924) Lepidoptera of New York and neighbouring States. *Cornell Univ. agric. Exp. Sta. Mem.*, **68** (1923): 729 pp.

FRACKER, J. B. (1915) The classification of Lepidopterous larvae. *Ill. biol. Monogr.*, **2**: 1–169.

HINTON, H. E. (1946) On the homology and nomenclature of the setae of Lepidopterous larvae, with some notes on the phylogeny of the Lepidoptera. *Trans. R. ent. Soc. Lond.*, **97**: 1–37.

MEYRICK, E. (1928) *Revised Handbook of British Lepidoptera*. London: Macmillan.

MOSHER, E. (1916) A classification of the Lepidoptera based on characters of the pupa. *Bull. Ill. Lab. nat. Hist.*, **12**: 15–159.

TILLYARD, R. J. (1923) On the mouthparts of the Micropterygoidea (Lep.). *Trans. ent. Soc. Lond.*, **1923**: 181–206.

TUTT, J. W. (1899–1909) *A Natural History of the British Lepidoptera*, 8 vols. (incomplete). London: Swan, Sonnenschein.

28. *Trichoptera*

MCLACHLAN, R. (1874–84) *A Monographic Revision and Synopsis of the Trichoptera of the European Fauna*. London: van Voorst.

MOSELY, M. E. and KIMMINS, D. E. (1953) *The Trichoptera (Caddis-flies) of Australia and New Zealand*. London: British Museum (Nat. Hist.).

ROSS, H. H. (1944) The Caddis-flies or Trichoptera of Illinois. *Bull. Ill., nat. Hist. Survey*, **23**, No. 1: 326 pp.

—— (1967) The Evolution and past dispersal of the Trichoptera. *A. Rev. Ent.*, **12**: 169–206.

29. *Hymenoptera*

BISCHOFF, H. (1927) *Biologie der Hymenopteren*. Berlin: Springer.

CLAUSEN, C. P. (1940) *Entomophagous Insects*. New York & London: McGraw-Hill.

EVANS, H. E. (1966) *The Comparative Ethology and Evolution of the Sand-wasps*. London & Cambridge, Mass: Comstock Publishing Associates.

FREE, J. B. and BUTLER, C. G. (1959) *Bumblebees*. London: Collins.

ISHAY, J. and IKAN, R. (1968) Food exchange between adults and larvae in *Vespa orientalis* F., *Anim. Behav.*, **16**: 298–303.

MICHENER, C. D. (1944) Comparative external morphology, phylogeny, and a classification of the bees (Hymenoptera). *Bull. Am. Mus. nat. Hist.*, **82**: 157–326.

SNODGRASS, R. E. (1956) *Anatomy of the Honey Bee*. Ithaca, N.Y.: Comstock Publishing Associates.

SUDD, J. H. (1967) *An Introduction to the Behaviour of Ants*. London: Arnold.

WILSON, E. O. (1971) *The Insect Societies*, Cambridge, Mass.: Harvard Univ. Press.

Phylogeny and Fossil Insects

ANDERSON, D. T. (1973) *Embryology and Phylogeny in Annelids and Arthropods*. London: Pergamon. 495 pp.

CARPENTER, F. M. (1977) Geological history and evolution of the insects. *Proc. 15th int. Congr. Ent.*, pp. 63–70.

CROWSON, R. A. *et al.* (1967) Arthropoda: Chelicerata, Pycnogonida, *Palaeoisopus*, Myriapoda and Insecta. In: *The Fossil Record* (Eds. Harland, W. B. *et al.*) London: Geological Society. pp. 499–534.

HENNIG, W. (1969) *Die Stammesgeschichte der Insekten*. Frankfurt a. M.: Kramer. 436 pp.

HINTON, H. E. (1958) The phylogeny of the Panorpoid orders. *A. Rev. Ent.*, **3**: 181–206.

IMMS, A. D. (1936) The ancestry of insects. *Trans. Soc. Brit. Ent.*, **3**: 1–32.

KRISTENSEN, N. P. (1975) The phylogeny of hexapod 'orders'. A critical review of recent accounts. *Z. zool. Evolutionsforsch.*, **13**: 1–44.

LAUTERBACH, K. E. (1973) Schlüsselereignisse in der Evolution der Stammgruppe der Euarthropoda. *Zool. Beitr.*, (N.F.) **19**: 251–99.

LAURENTIAUX, D. (1953) Classe des insectes. In: *Traité de Paléontologie* (Ed. Piveteau, J.). **3**: 397–527.

MANTON, S. M. (1973) Arthropod phylogeny – a modern synthesis. *J. Zool. Lond.*, **171**: 111–30.

—— (1977) *The Arthropoda*. Oxford University Press. 527 pp.

SNODGRASS, R. E. (1952) *A Textbook of Arthropod Anatomy*. Ithaca, N.Y.: Comstock Publishing Associates. 363 pp.

TUXEN, S. L. (1970) The systematic position of entognathous Apterygotes. *An. Esc. nac. Cienc. biol. Méx.*, **17**: 65–79.

Index

Names of authors are in small capitals. Generic names are in italics. Synonyms are indicated by a cross-reference (e.g., *Calandra*, see *Sitophilus*). Page numbers in bold type denote illustrations.

Planococcus, 174
planta, 28
Plant-lice, see Aphidoidea
plantulae, 27
plasma, 90
plastron respiration, 87
Platycnemis, embryology of, 112, 114
Platyedra, 199
Platygaster, polyembryony, 108
Plecoptera, ancestry of, 156; characters
  of, 145, 156, 157, 158; instar number,
  121; literature on, 228; ovipositor
  absent, 42
pleometrotic colonies, 138
Plesiocoris, 170, 171; nymph, 171
pleural apophysis, 35
pleural suture, 25, 26
pleural wing-process, 26
pleurite, 20, 25, 26
pleuron, 20
pleurosternal muscle, 34, 35
Plusia, see Autographa
Podura, 132
poison canal, 42
poison gland, 42, 97
poison sac, 41, 42
POISSON, R., 233
polarized light, perception of, 69
pollen basket, see corbicula
polyembryony, 107, 108, 206
Polymastigina, 165
Polymitarcys, nymph, 153
polymorphism, in ants, 140; in aphids,
  172; in termites, 164; in locusts, 159
Polyphaga, 185
polypod embryo, 116
polypod larva, 124
polytrophic ovariole, 101
Pomace-fly, see Drosophilidae
Pontomyia, 131
POPHAM, E. J., 229
pore-canal, 15, 16
postantenatal organ, 150, 151
posterior tentorial pit, 21
postmentum, 23
postnotum, 25, 26
post-occiput, 21
postoesophageal commissure, 54
poststernellum, 25, 26
POTTER, E. M., 235
predatory insects, 134
prementum, 23; in Diptera, 186, 187
pre-oral food cavity, 70
prepupa, 125
prescutum, 25, 26
presternum, 25
prestomal teeth, 186, 187, 190
prestomium, 186
pretarsus, 26, 28
PRIESNER, H., 233

primary host, of aphids, 172
PRINCIS, K., 230
PRINGLE, J. W. S., 224
priority in nomenclature, 143
Pristiphora, 207
proboscis, of Diptera, 187, 190; of
  Hemiptera, 169, 171; of Lepidoptera,
  195, 196
procephalic lobes, 116
proctodaeum, 69, 113, 117
Proctotrupoids, 209; larval, 124
prognathous head, 21
prohaemocyte, 90, 91
Projapygidae, 149
propneustic tracheal system, 83
propodeum, 202, 205, 207
proprioceptor, 57; intramuscular, 59
Prorhinotermes, 165
prostheca, 23
protease, 75, 119
prothoracic gland, see thoracic gland
prothorax, 24
protocephalic region, 116
protocerebrum, 53
protocormic region, 116
Protodonata, 218
protopod embryo, 116
protopod larva, 123, 124
Protozoa, in Isoptera, 145
Protura, characters of, 144, 149, 150,
  217; literature on, 226
proventriculus, see gizzard
Pseudococcus, 174; female, 173; head,
  171
pseudotracheae, 186, 187, 188
Psocoptera, characters of, 145, 166;
  literature on, 231; ancestral to
  Hemipteroids, 219
pterines, 18
Pterophoridae, 200
pterostigma, 32
Pterygota, 144; earliest fossils, 215
Pthirus, 168
ptilinal suture, 191
ptilinum, 191, 192, 194
Pulex, length of jump, 183-4
pulsatile organs, 90
pulvillus, 27, 28
pupa, 123, 125, 126; of aberrant Exopterygota, 123; in Lepidoptera, 199;
  types of, 126, 127
pupal cell, 125
puparium, 126, 127, 192, 193
pyloric sphincter, 74
Pyralidae, 200
Pyrausta, see Ostrinia

Quadraspidiotus, 174
quadrate plate, 40, 41
queen bee, 139